Climate and Development

Edited by
ASIT K. BISWAS

TYCOOLY INTERNATIONAL PUBLISHING LIMITED
DUBLIN

First published 1984 by
Tycooly International Publishing Limited
6 Crofton Terrace, Dun Laoghaire, Co. Dublin

ISBN 0 907567 36 3 (Library edition)
ISBN 0 907567 37 1 (Softcover)

Typeset by Photoset Ltd., Dublin
Printed in Ireland by Mount Salus Press Ltd., Dublin 4.

British Library Cataloguing in Publication Data

Climate and development — (Natural resources
and the environment series; v. 13)
1. Economic development 2. Man — Influence of climate
I. Biswas, Asit K. II. Series
330.9 HD75

Contents

Preface

The interrelatedness of climate and development was clearly illustrated during the early seventies. The aggregation of the climatic events that contributed to the prolonged drought in the Sudano-Sahelian region of Africa, the failure of the Russian grain harvest, the erratic monsoons in the Indian subcontinent, the disappearance of the anchovy fishery off the coast of Peru, and the serious drought conditions in the western part of North America, was a global catastrophe of major magnitude. The gravity of the condition that developed in 1972 can be best illustrated by the fact that the total world output of food declined from the preceding year for the first time in twenty years, due to the adverse weather conditions. The production of cereals — wheat, coarse grains and rice — which form the staple element of diet for most of mankind — declined by 33 million tons instead of an anticipated increase of 25 milion tons. This created a serious world food problem, especially as two of the main food-exporting countries, the United States and Canada, had instituted policies to reduce their large surpluses. Consequently, surplus wheat stock in exporting countries fell from 39 million tons in 1971-72 to 29 million tons in 1972-73, and even further in 1973-74. Rice reserves were virtually exhausted.

Both shortage of cereals and their high world market price created a serious food problem for the developing world. Under such critical conditions, the United Nations General Assembly, in December 1973, decided to convene the World Food Conference at the highest decision-making level. Initially proposed by Henry Kissinger, the then US Secretary of State, the World Food Conference was held in Rome, Italy, in November 1974.

The serious food situation of the early seventies clearly indicated that the opinion expressed by some scientists and decision-makers in the sixties that technological developments have made agricultural production independent of the vagaries of climate was both misplaced and optimistic.

While the interrelatedness of climate and development is self-evident, for various reasons which I have discussed in detail in our earlier book *Food, Climate and Man* (John Wiley & Sons, New York, 1979), development planners on the one hand have seldom considered climate as an explicit factor in the development process and climatologists on the other hand have not generally played an active part in development planning. This situation needs to be rectified as soon as possible in order that the development process can be made sustainable on a long term basis.

For the developing countries of the world, which are located in the tropics and sub-tropics, climate should be considered to be an important resource, which provides certain golden opportunities for development but also simultaneously poses some constraints. Hence, development strategies have to be formulated that specifically attempt to maximize the benefits such opportunities can bring but do not forget the constraints imposed by climate. Technically, such a task is not going to be easy. Furthermore, direct technology transfer from temperate to tropical regions may not be the optimal solution, and in many cases may even be counterproductive.

During the next two decades, the population of tropical countries is expected to increase by 1.5 billion, which will account for approximately 90 per cent of the anticipated global population increase. Provision of basic human needs to these unborn generations, and improvement of the quality of life of the present underprivileged people on this earth, is a critical challenge. If past experience is any indication, it will not be easy.

The task ahead of us is a most challenging and complex one, and it can only be accomplished if we proceed in a scientific and rational way. It is essential that we must draw upon whatever knowledge we have gathered on the tropical areas up to now, and then supplement this information base with as much additional knowledge as is practically possible in order to formulate appropriate strategies for sustainable development.

One luxury we do not have is time. We cannot wait until all the relevant information is available before making development decisions. People are already here on earth, and millions are going to be born before the end of this century, whose needs have to be met from the earth's resources. Herein lies one of the great challenges of the future: how to make the right development decisions despite all the risks and uncertainties involved.

Climate and development are interrelated in many ways, some of which were discussed in our earlier book *Food, Climate and Man.* The main focus of *Climate and Development* is on issues that are of immediate concern to people, especially for those in developing countries of tropics and semi-tropics. While some mention has been made of the climate-development related problems that may arise in the twenty-first century, like carbon dioxide or changes in precipitation and temperature regimes, these have not been discussed in depth. Instead our emphasis has been on the impacts of climate on terrestrial biota, especially agricultural development, on water resources planning and management and human health, and climatic risk assessment through input-output models.

Like *Food, Climate and Man,* the present book owes much to Dr Mostafa Kamal Tolba, Executive Director of the United Nations Environment Programme. Dr Tolba not only encouraged me to undertake the preparation of this book, but also — in spite of his manifold commitments all over the world — generously gave me his time to discuss aspects of the issue, whenever necessary. Without Dr Tolba's encouragement, it would not have been possible to complete this book. For this I am truly grateful.

Asit K. Biswas

March 1984
Oxford, England.

CHAPTER ONE

Climate and Development

Asit K. Biswas
International Society for Ecological Modelling

DURING THE LAST three decades, differences between expectations of development patterns and the developments that have taken place are indeed remarkable. As expected, the development patterns of the past three decades have varied not only from region to region, but also within individual countries. They have also varied with time depending on the internal problems and external constraints encountered. There is no reason to doubt that future development trends will be similarly different from what are anticipated due to lack of reliable methodological bases for forecasting as well as unforeseen and unexpected events.

During the past three decades our perception of development has also changed. In an excellent analysis of the changing pattern of perception of development problems, Misra (1981) observed:

> Just thirty years ago, all those who mattered in development — politicians, academics and planners — appeared to know what development meant and how to achieve it. There was a mood of confidence, assurance and urgency. There were instant solutions — decolonization, economic aid, industrialization, etc. There were theories and strategies galore — often contradicting each other, yet accepted ungrudgingly as they were the products of some of the brilliant minds of the time. . . .
>
> Today, thirty years later, the success euphoria of the past has given place to despondency, confusion and stalemate. The developers are still there, and so are the development theorists and planners. But they all concede that development is not as easy as they thought it to be and that there is no panacea for underdevelopment. It is realized that the road to development is tortuous; that development is not economic growth alone, and that many of the issues which were debated in the past, were not the real issues when seen in the light of the problems being faced by the less developed countries today.

Many new nations became independent during the fifties, sixties and seventies, and they made their own plans for future development. Amidst the euphoria of independence, there was hope for the future, and expectations of "good" life for all their citizens. In some cases the dreams have been shattered; many people are now worse off than they were before, and the rosy future they all expected to see in not too distant a timeframe — certainly within their life-time — has retreated even further and further like a mirage. Disillusion and despondency have replaced hope and euphoria.

The national expectations of the recent decades had international impact as well. Development goals and targets were agreed to for both the First and the Second Development Decades of the United Nations covering the sixties and seventies. These targets were unfortunately not achieved. Bradford Morse (1980), Administrator of the United Nations Development Programme, reviewing the achievements of the seventies, said:

> From one point of view, the 1970s was a decade of disappointments. Adequate gains were not made against poverty and its life-crushing consequences. The global economy fell short of the sustained expansion necessary for moving with much greater speed and effectiveness in the struggle to substantially ease hunger, disease, illiteracy, unemployment and lack of adequate housing. The world became joltingly aware that there were limits to its exploitable resources. Perhaps most frustrating of all was the fact that the industrialised and the developing countries did not achieve greater understanding — much less agreement — about how to deal with these problems effectively and equitably.

Few people will disagree with the above assessment. Similarly, during the Third General Conference of the United Nations Industrial Development Organisation (UNIDO, 1980), held in New Delhi in early 1980, many governments expressed the opinion that the "two United Nations Development Decades had failed in their objectives". The good intentions and objectives of UNIDO's Lima Declaration and Plan of Action of 1975, which stipulated that the developing countries should attain a 25 per cent share of the total world manufacturing output by the year 2000, appears to recede further on the horizon. At the present rate of growth, their share might not exceed 13 per cent by the end of the present century — a figure that is only half the accepted target. The conference stated that during the two decades, "the rich became richer and the poor poorer; more than one-quarter of the world's population was growing steadily poorer." Further, "eight hundred million people, or about 40 per cent of the population of the developing countries continued to live in absolute poverty; roughly a billion people lacked at least one of the basic necessities of food, water, shelter, education or health care."

Agriculture dominates the economies of most developing countries. For low-income economies, nearly 70 per cent of the population are directly dependent on agriculture to earn their livelihood, either as farmers or farm workers. It is not uncommon to find people who have to spend 60 to 75 per cent of their income on food-related purposes. The role of agriculture in overall development of developing countries should not be underestimated. Agricultural products are still the main category of export for most developing countries, and accounted for some 30 per cent of their total export earnings in the late seventies. Experience clearly indicates that those developing countries that embarked on industrialization at the expense of the agricultural sector have not fared well. The latest World Development Report (World Bank, 1982) states categorically that "one point emerges very clearly from the diversity of experience of the developing countries: rapid growth of agriculture and in GDP go together." It has become increasingly evident during the past three decades that sustained development in most developing countries is unlikely to occur without first or simultaneously developing their agricultural potential.

Since agriculture has a major impact in developing countries, at least initially in the development process, and since agricultural production is still closely related to climate, progress in the agricultural sector in the recent past is an important factor to consider for any discussion of the interrelationships between climate and development.

The target for the average annual growth rate for the agricultural sector during the Second Development Decade was established at 4 per cent which, if achieved, would have been comfortably ahead of the rate of population growth. The real average annual growth rate was, however, only 2.8 per cent. It should be noted that this was the average growth rate: it varied remarkably from one country to another. Several countries were significantly worse off. If the developing countries are considered as a whole, certain indicators of agricultural production during the first two Development Decades actually declined. For example, developing countries, in aggregate, were net exporters of grain in the 1950s. At the end of the First Development Decade, the surplus situation had turned into a net deficit. Developing countries as a whole imported 42 million tons of grain in 1970, and this further increased to 80 million tons by 1979. Estimates of total grain import needs by the end of the Third Development Decade in 1990 currently range from 125 to 150 million tons (Biswas and Biswas, 1981).

Similarly, if the index of per capita food production is considered, the situation is not much better for the low income developing countries, defined by the World Bank (1982) as having gross national product (GNP) per capita of US$410 or below in 1980. Of the 33 such countries listed in the World Bank's World Development Report of 1982, the index of per capita food production (1969–71 = 100) declined in 1978–80 for twenty-three countries, remained the same for one, and increased for only nine countries. The index declined to a low of 41 for Kampuchea, 75 for Mozambique and 81 for Togo and increased to a high of 121 for Sri Lanka, 116 for China and 107 for Vietnam. Some of the development indicators of selected countries are shown in Table 1.1.

There is no doubt that the progress during the first two development decades in the agricultural sector has not matched with expectations. Maurice J. Williams, Executive Director of the World Food Council, called it twenty years of neglect when describing the results of the two decades on the agricultural sector in the Third World. According to Williams:

> The disturbing features of the longer term trends relate, in particular, to the inadequate rate of increase in food and agricultural production in the developing countries, the continuing rise in their food imports, the deterioration in their food self-sufficiency, lack of evidence of any reduction in the incidence of hunger and malnutrition, slow progress in the establishment of an effective system of world food security, decline in the share of developing countries in agricultural export earnings and inadequate flow of external assistance. The decade of the eighties is thus starting with a heavy backlog of unresolved food and agricultural problems.

One major problem has been due to the fact that many developing countries did not give the agricultural sector the necessary priority in their national development plans. Fortunately, this viewpoint, which usually fosters heavy industry at the expense of agriculture, appears to be changing. The recommendation of the Group of 77 at the

Table 1.1. Development indicators of selected countries (Source: World Bank, 1982)

Country	Population (10^6) mid-1980	GNP/capita (US$) 1980	Adult literacy (%) 1977	Life expectancy at birth (yrs) 1980	Index of food production/capita 1978-80	Commercial energy consumption (kg coal equivalent) 1979
Argentina	27.7	2,390	93	70	122	1,965
Bangladesh	88.5	130	26	46	94	40
Brazil	118.7	2,050	76	63	117	1,018
China	976.7	—	66	52	92	734
Egypt	39.8	580	44	57	93	539
France	53.5	11,730	99	74	115	4,810
Germany, FR	60.9	13,590	99	73	110	6,264
India	673.2	240	36	52	101	194
Indonesia	146.6	430	62	53	110	225
Japan	116.8	9.890	99	76	93	4.048
Kenya	15.9	420	50	55	86	172
Korea, R.	38.2	1,520	93	65	130	1,473
Mexico	69.8	2,090	81	65	103	1,535
Singapore	2.4	4,430	—	72	147	5,784
USSR	265.5	4,550	100	71	108	5,793
UK	55.9	7,920	99	73	118	5,272
USA	227.7	11,360	99	74	115	11,681

United Nations, which includes all the developing countries, during the finalisation of the strategy for the Third Development Decade, was quite unequivocal. The group called for "a distinct and definite bias in favour of agricultural production," and an average annual growth rate of 4 per cent. They further suggested that food and nutritional planning should form the core of national development policies. This is a positive indication, since the recommendation comes from the group of countries that will be most affected. While this, without any question, is a step in the right direction, it remains to be seen whether such resolutions are rhetorical or will actually be implemented in the countries concerned.

The above discussion, however, should not be taken to imply that developments during the past three decades have been all negative, but rather that expectations were not fulfilled. On the positive side, technological developments and improved management practices increased agricultural output at approximately twice the rate of earlier periods. But rapid population growth, skewed income distribution and changing patterns of growth often tended to exacerbate the overall problem. For example, the world population increased from 2.8 billion in 1955 to 4.4 billion in 1980, a 57 per cent increase in only 25 years. While in South Asia the balance between population growth rates and agricultural growth rates (2.5 per cent and 2.2 per cent respectively) could be maintained during the past two decades, the situation deteriorated significantly in Africa, where agricultural growth rates declined from 2.7 per cent in the sixties to only 1.7 per cent in the seventies. The situation has worsened since the rate of growth of population has accelerated.

The extent and magnitude of unequal income distribution between the masses — both within a country and between countries — is an area of major international concern at present. It is not unusual to find 20 per cent of the people in the highest income bracket of a country account for 50 to 70 per cent of GNP of that country, but the lowest 20 per cent contribute only 5 per cent of GNP. This situation is also generally valid when land ownership is considered. According to FAO statistics, 20 per cent of the richest landowners own 82 per cent of cropland in Venezuela, 56 per cent in Colombia, 53 per cent in Brazil and 50 per cent in India, Pakistan and the Philippines. Since the rich control the political power, not unexpectedly the vested interests would like to maintain the *status quo*.[1] When the two phenomena of increasing population growth and unequal income distribution are considered simultaneously, it means that millions of people have not only sharply reduced food available per capita — compared to earlier periods — but also do not have access to other benefits that could accrue from increasing GNP per capita, that has occurred in the vast majority of developing countries.

Changing patterns of growth is also another concern. According to the World Bank (1982) some 600 million tons of cereals are fed to animals every year at present, enough to feed 2.5 billion people, or more than double the present number in poverty. Furthermore, the efficiency of conversion of cereals to meat is very low: 75 to 90 per cent of calories and 65 to 90 per cent of protein is lost in the process. This, however, does not mean that changing feed grain to food will resolve the world hunger problem.

1. This is exemplified by one of the author's recent experiences when advising an important country to prepare a national agricultural development plan. One of the specific instructions was *not* to consider the problem of land tenure.

Since much of the feed grain used is in developed countries, the production will decline sharply if the market for feedstock is restricted. Until a reliable market for the cereals can be created in developed countries, which means significantly increasing their purchasing powers, the problem is unlikely to be resolved.

CLIMATE AND DEVELOPMENT

It can be legitimately asked why, despite international concern and a multitude of national efforts, agricultural performances in the majority of developing countries have not matched general expectations. This is a difficult and complex question to answer, since many factors are involved and the various issues concerned are mostly interrelated. However, one of the important reasons for such a poor performance in the past has to be the sad neglect of the impacts of climate on the development process itself. As the Nobel-Laureate Gunnar Myrdal (1968), in his monumental work, *Asian Drama* has aptly noted, economic analysis has tended to disregard climate, except occasionally in the location theory, in spite of the fact that "climate exerts everywhere a powerful influence on all forms of life — vegetables, microbial, animal and human — and on inanimate matter as well." It is a sad but true fact that economists, including agricultural and development economists, have consistently ignored climate. Almost all macro- and micro-economic growth models fail to explicitly consider climate as a major parameter. Similarly, climate is seldom discussed in books dealing with regional development theories.

This situation is somewhat surprising, since the impacts of climate on development have been realized for at least 2,400 years, albeit not fully or comprehensively. For example, Hippocrates (460-400? BC), father of medicine, attested:

> I hold that Asia (Minor) differs very widely from Europe in the nature of all its inhabitants and all of its vegetation. For everything in Asia grows to far greater beauty and size; the one region is less wild than the other, the character of the inhabitants is milder and more gentle. The cause of this is the temperate climate, because it lies towards the east midway between the risings of the sun, further away than is Europe from cold.

Very few development specialists in recent years have realized the importance of considering climatic factors in the development of nations. In 1951, Galbraith noted that if "one marks off a belt a couple of thousand miles in width encircling the earth at the equator, one finds within it no developed countries . . . . Everywhere the standard of living is low and the span of human life is short." A decade later, a United Nations (1961) report noted that if "the industrialized countries are marked on a map, they will be seen to be located as a rule in colder climate than the underdeveloped countries. The correlation with climate is as good as most correlations between non-economic factors and economic development." Similarly Myrdal (1968) has urged that every "serious study of the problems of underdevelopment and development in the countries of South Asia should take into account the climate and its impacts on soil, vegetation, animals, humans and physical assets — in short, on living conditions in economic development."

Few other economists have, however, dismissed the role of climate in development. Thus, Lee (1957) stated: "Climate and economic development in the tropics is a convenient bogeyman to be blamed for psychological difficulties whose real origin is much more personal." Similarly, according to Lewis (1955): "Because economic growth is currently most rapid in temperate zones, it is fashionable to assert that economic growth requires a temperate climate, but the association between growth and temperate climate is a very recent phenomenon[2] in human history." Lewis further noted that the "climate hypothesis does not take us very far."

One can, however, ask if development economists have failed miserably to consider climate as an important factor for development planning, why have not the climatologists ensured that such a neglect is not allowed to continue. The answer is fairly simple. Much though the climatologists know about climate, they have not ventured out of their own discipline: they have tended to remain isolated within their own field – a phenomenon that is very common for nearly all professions. Very few climatologists have ventured into such fringe areas as development planning. True, there have been some recent attempts to develop a better fundamental framework of theory in this area (i.e. Kamarck, 1976), but even this has been largely ignored. Accordingly one is indeed hard pressed to name more than a handful of climatologists who are even active in the fringe areas of development. It is a sad but true fact that most internationally acknowledged development planners will find it difficult to name even one climatologist whom they would include within their peer group. Accordingly it is no surprise that an analysis of the proceedings of the World Climate Conference (1979), a "conference of experts on climate and mankind", held under the aegis of the World Meteorological Organization (WMO) of the United Nations, does not contain a single paper that analysed the relations between climate and overall development. It should, however, be noted that in the past often the study of the interrelationships between climate and living organisms was primarily in the domain of physiologists (Tosi, 1975).

DIFFERENT CONDITIONS FOR DEVELOPMENT

While it has been argued earlier that climate has important impacts on development patterns, it should not be construed to mean that the interrelation between them is directly proportional. Nor does it mean that just because economic development has occurred in recent decades in the temperate regions rather than in the tropics and semi-tropics, the regions underdeveloped will continue to remain so in perpetuity. In other words, no sensible person will subscribe to the glib doctrine of pessimistic geographical determinism that was prevalent during the colonial era which explained the poverty of developing countries in terms of their climates. According to this widely

2. In refuting this type of statement, Myrdal (1968) observes that the great civilizations that sprang up in tropical areas in ancient times and that lasted for centuries are different "in fundamental ways from modern ones; also, they often grew up in smaller regions favoured by exceptional conditions; soil erosion and deforestation had not proceeded so far and so on."

believed theory of the time, the economic underdevelopment of the Third World was due to its "unbearable" climate and its impacts on the different components of the biosphere, both living and inert. It indirectly reinforced the concept of the racial inferiority of the colonized people, and according to Myrdal (1968), such a concept of geographical pessimism '"supported the common view, badly needed as a rationalization of Western colonial policy, that little could be done to improve the productivity of the colonies and the life of the colonial people." Reviews of the theories of geographical determinism have been made by Biswas (1979), Kamarck (1976) and Myrdal (1968).

A rigorous analysis of countries in the temperate regions and in the tropics will clearly indicate that the boundary conditions for development for the two regions are not identical: in fact they are very different. Since the boundary conditions, within the context of which economic development takes place, are different in the temperate and tropical regions, it means that the patterns of development experienced in the Western industrialized countries may not be duplicated in the Third World. Myrdal points out that it is "important to take note of the newness of the development problems confronting the countries of South Asia today — and most other underdeveloped countries — because of the tendency to overlook their uniqueness that is inherent in the biases common in research and prevalent also in planning and, generally, in public discussion."

Similar misgivings on differing boundary conditions have been expressed by Streeten (1971): "the deep-seated optimistic bias with which we approach problems of development and the reluctance to admit the vast differences in initial conditions with which today's poor countries are faced compared with the pre-industrial phase of more advanced countries." Similarly, Kamarck (1979) suggested that the proper contrast is not "north-south" but "rich temperate zone-poor tropics".

The different boundary conditions in the tropics and temperate regions are not necessarily reflected by existing widely-accepted development theories or development processes. Boulding (1970) has remarked:

> Development, like economics, has been very largely a temperate zone product. The complexities both of tropical ecology and of tropical societies are beyond easy access for those raised in essentially temperate zone culture. This is not to suggest a naive climatological determinism, but just as tropical biological ecosystems differ very markedly from those in the temperate zone, it would not be unreasonable to suppose that the processes of social evolution would likewise produce marked adaptations to the peculiar rigors and delights of tropical climate and life style.

One can argue that the general failure of development planners and scientists to recognize the importance of climate in the development process may be considered to be another tragic facet that has contributed to numerous instances of failures due to technology transfers between the temperate and tropical regions. Attempts to linearly transfer both development concepts and technologies that originated in the temperate countries to the tropics have often not been successful, and in some instances they have only been partially successful, with highly reduced effectiveness. The agricultural history of the present century is replete with examples in which straightforward transfer of technology from one region to another actually created additional problems. A few select examples are the deep-ploughing of the rice paddies in Java by the Dutch, corresponding operations by the British in Burma, failure of the groundnut scheme in Tanzania, broiler production in Gambia and the folly of cultivating marginal lands which should never have been farmed in many African, Asian and Latin American countries.

Probably the most spectacular failure in agricultural development was the 1947 British plan to develop large-scale groundnut plantations in East Africa, in what was then known as Tanganyika. The area selected covered 3.25 million acres, 70 per cent of which were uninhabited, for what later turned out to be good reasons. All sorts of experts were recruited for the ambitious project, but overall planning left much to be desired. Two of the three areas selected turned out to be unsuitable for cultivation. At Kongwa, the precipitation was too low and the soil was very compact and abrasive. Similarly, at Urambo, much of the land was low-lying and hence subject to waterlogging (Young, 1976). Some of the practices were not environmentally sustainable. For example, bulldozers were extensively used to remove deep tree roots. The soil, as in several other similar cases in the tropics, could not stand up to the machines, and there were severe losses due to wind and rain. Artificial fertilizers were not effective because of lack of water, and germination turned out to be difficult in hard-packed soil. The project was eventually abandoned after six years of desperate efforts and capital investment of some US$100 million. It was a classic example of an attempt to develop a large area without adequate soil and natural resources surveys.

The failures of many agricultural development schemes in the tropics, even though they used well-proven and workable models from temperate regions, should not come as a surprise. Most of the theories in economic development or economic geography are products of the Western world, and their fundamental principles, evolved over the years, are generally based on conditions prevalent in developed countries. Thus, many "classical" theories are being applied in the tropics, even though they are primarily temperate zone products, and thus may be of questionable validity. When these suspect theories are superimposed on a different world, on an alien culture with vastly different socio-economic conditions, religious-cultural practices and institutional infrastructures, the risk of committing a fundamental error is exceedingly high. If, for example, the underutilized labour force, a common condition in Asia, Africa and Latin America, is analysed according to traditional Western concepts of unemployment and under-employment, the resulting figures and conclusions are generally meaningless, or at best, the magnitude of error is so great it would be folly to rely on them to formulate major policy decisions. To quote Myrdal (1968) again:

> The very concepts used in their (theories of classical economics) construction aspire to a universal applicability they do not in fact possess. As long as their use is restricted to our part of the world this pretence of (universal) generality may do little harm. But when theories and concepts designed to fit the special conditions of the Western world — and thus containing the implicit assumptions about social reality by which this fitting was accomplished — are used in the study of underdeveloped countries in South Asia, when they do *not* fit, the consequences are serious.

Since development theories from the Western industrialized countries generally do not fit the tropics, why are not appropriate theories being formulated by scientists from the developing countries themselves? Herein lies one of the great dilemmas of the modern times. At present most of the elite in the developing world tend to be trained in the West, and in general, Western thinking is considered to be more "progressive", "advanced" or "scientific". Because of such training and social attitudes, these intellectuals usually produce dissertations replete with the traditional theories of classical Western economics. Many are familiar with the latest abstract growth models originating from Harvard or Oxford, but very few question the validity of their use within the context of differing socio-economic and institutional conditions in their own countries. This means the underlying biases go undetected and are perpetuated, when at the very least they should be identified and questioned, and better still corrected. In addition, such uncritical acceptance of the bias in Western concepts and theories on the part of academia is not confined to any specific country or to a regional grouping of countries: it seems to permeate through all countries having similar economic systems. Fortunately, it appears that this situation is beginning to change: increasingly a few are questioning their validity and usefulness.

Climate has impacts on many different aspects of development. Only some of the important aspects will be discussed herein.

TROPICAL CLIMATES: THE DIFFERENCE

During the happier days of the sixties, many people claimed that technological developments had freed modern agriculture from the vagaries of climate. To some extent, such overconfidence can be accounted for by the generally benign nature of the climate in the sixties. It became clear during the early seventies that climate still was a major factor for overall agricultural production, and that earlier technological overconfidence was highly misplaced.

Since both the climatic and environmental conditions between the tropics and temperate regions are different, the agricultural practices of the industrialized countries cannot be directly duplicated in developing countries. The difference in boundary conditions on factors that affect land, which makes direct technology transfer process hazardous, is worth analysing.

The two most important climatological factors that affect agricultural production are rainfall and temperature. The nature and distribution of rainfall, both within a year and from year to year, tend to be different in the tropics when compared to temperate locations. Table 1.2 shows the average monthly rainfall in millimetres at London, Sokoto on the southern border of the Sahel, and at Lagos, Nigeria, representing typically equatorial rainfall (Ormerod, 1978). The yearly average rainfall for London and Sokoto do not differ appreciably: 568 mm and 668 mm respectively. However, when distribution of rainfall throughout the year is concerned, the two cases are very dissimilar. The rainfall pattern of London, which has a temperate climate, can be characterized by a low but reasonably uniform monthly rate over the entire year. It varies from a maximum of 61 mm in October to a minimum of 35 mm in April. Similarly rainfall retained in the soil is reasonably uniform. The situation is very different for Sokoto, where the rainfall is intense during July to September, but virtually non-existent between October to April. The rainfall varies from a maximum of 239 mm in August to zero between November to March. Furthermore, Sokoto has a significantly lower rainfall retention rate in the soil when compared to London. Thus, even though the total average annual rainfall in Sokoto is actually 15 per cent higher than in London, its distribution throughout the year is very

Table 1.2. Average monthly rainfall in millimetres. Bracketed figures are for rainfall retained in soil (Source: Ormerod, 1978)

Month	London	Sokoto	Lagos
January	41 (245)	0 (25)	26 (100)
February	37 (276)	0 (16)	46 (73)
March	41 (290)	0 (9)	100 (60)
April	35 (281)	10 (5)	148 (60)
May	41 (253)	48 (3)	269 (182)
June	48 (213)	86 (2)	457 (300)
July	56 (175)	147 (2)	273 (300)
August	58 (152)	239 (104)	64 (250)
September	42 (138)	145 (107)	138 (266)
October	61 (156)	13 (67)	205 (300)
November	54 (187)	0 (42)	68 (230)
December	54 (226)	0 (31)	26 (150)
Total	568	668	1,820

uneven, making Sokoto very arid. The annual rainfall in Lagos is very high, 1,820 mm, but this occurs during two parts of the year, long rains between March to July and short rains during September and October. The rainfall is very low between November to February and again in August. The maximum rainfall is in the month of June, 457 mm, which represents nearly 80 per cent of the annual average rainfall of London, and the minimum is 26 mm for the months of December and January. The monthly distribution of rainfall retained in the soil varies from a minimum of 60 mm to a maximum of 300 mm — a factor of 5. For London, the identical ratio is only 2:1.

Another feature worth noting from Table 1.2 is the relatively poor moisture retention capacity in the two tropical stations, which is likely to be primarily due to the difference in organic content of the soil. Generally, organic content of soil comprises several compounds of the humic and fulvic acid types, which are formed by microorganisms in the breaking down of cellulose (Schnitzer, 1976). While there is no fundamental difference between tropical and temperate soils in terms of the characteristics of these compounds, remarkable differences often exist in terms of their overall content (Schnitzer, 1977). These compounds are important for soil because of their ability to retain water and mineral salt and resistance to leaching. Ormerod (1978) points out that "high temperatures, long periods of drought, intense ultraviolet radiation and particularly high kinetic energy rainfall, which destroys the granular structure of the soil, decrease the activity of soil microorganisms so that there is little possibility in open land for the stable organic content of the soil to build up; indeed there is a tendency for it to be destroyed." This, however, does not mean that high organic content cannot be built up in tropical soils under certain specific conditions, a point which will be discussed later.

The variation in rainfall in tropical climates often tends to be greater than in temperate regions. For example, in Pakistan, total annual rainfall in any given year can be expected to exceed or fall short of the mean annual rainfall by an average of 30 per cent or more (Stamp, 1966). This type of erratic rainfall is problematical from an agricultural viewpoint, since the rainfall could be either too much or too little, thus requiring extensive and expensive water control systems for irrigation,[3] drainage and flood control.

In the monsoon countries, the timing of the onset of the monsoon is vitally important. Many a famine in the Indian sub-continent is directly due to the fact that the monsoon rains did not start at the right time. Equally important is the continuation of the rain once it has started. There have been several years when the onset of the monsoon rains were at the appropriate time but failed soon thereafter causing agricultural havoc. For the areas of Asia cultivating wet rice with rainfed farming, both the onset and continuation of the monsoons are vitally important factors. The importance of wet rice can be realised by the fact that it currently supports the highest density of people in the humid tropics, especially in Asia.

Rainfall has a direct impact on soil erosion all over the world, but the potential ability of the tropical rainstorms in causing soil erosion is far higher than in the temperate regions. This could be attributed to the high kinetic energy of the tropical rainstorms when compared to the gentler kinetic energy of rainfall in temperate regions. Kinetic energy of rainfall depends on the size of drops, intensity and wind velocity. While long-term detailed data on tropical rainfall are not available, it appears that median drop size of well above 3 mm are not uncommon. Drop sizes as high as 4.9 mm have been observed. From these data, a preliminary observation could be that the drop-size distributions of rainstorms is much higher in the tropics than in temperate regions.

Kinetic energy of rainfall is an important consideration since kinetic energy and the impact of raindrops initiates loosening and detachment of soil particles, the first essential step for soil erosion. Once soil is loosened, the particles are washed away, thus contributing to serious soil erosion problems.

So far as intensity of rainfall is concerned, it appears that its erosive power significantly increases at about 35 mm h^{-1}, which can be considered to be a threshold for erosion. Since more rainstorms in the tropics equal or exceed the level of this erosive threshold, the erosive potential is higher in the tropics when compared to the temperate parts (Hudson, 1971).

Another climatic aspect further contributes to soil erosion. The rainfall and temperature distribution patterns in tropical climates, especially in those areas having pronounced dry and wet seasons, the soil erosion problem is accentuated. During the long dry season, there is some loss of topsoil due to wind erosion. However, far more damage is done during the onset of the rainy season. The vegetative cover, at the end of the dry season, is already reduced and often at an absolute minimum. Thus when a heavy thunder shower occurs, the water does not infiltrate into the soil as it might in light steady rain, and year after year soil erosion takes place due to surface runoff. Tempany and Grist (1958) have suggested that if the heavy rains double the water flow, "scouring capacity is increased four times, carrying capacity thirty-two times and the size of particles carried sixty-four times." Fisher (1961) estimates that these processes have contributed to the erosion of nearly 150 million acres in India alone. Even considering the fact that soil is

3. It is interesting to note that even though much food is produced in the United Kingdom there is very little irrigation.

formed more quickly in the tropical region than in temperate climates — Veleger, according to Fisher (1961) estimates it to be ten times faster in the tropics — the soil formation is much too slow to replace the loss.

The above discussion does not mean to imply that tropical soils cannot be protected from severe erosion. By appropriate land use patterns and management techniques, it should be possible to reduce and control the erosion problem significantly. The overriding fact, however, remains; we still do not know enough about tropical soil taxonomy. Furthermore, since most of the research on soil taxonomy has occurred in the West — where tropical or sub-tropical regions generally do not exist — climatic information on many tropical soils remain unknown (Vehara 1981). Under these circumstances technology transfer is not an easy process.

Tropical forests and woodlands provide an interesting case. They are stable because over long periods of evolution, spanning the geological time scale, they have developed resilience which allows them to withstand climatic and other natural environmental hazards. However, faced with modern development and technology, the ecosystems may prove to be quite vulnerable. The ecology of tropical forests has to be much better understood before any long-term sustainable development plans can be made with any degree of confidence.

Tropical vegetations often give a deceptive impression of soil fertility. Major tropical forests often grow on nutrient-poor soils, especially in terms of phosphorus and potassium. During their evolutionary process, they have become adapted to such poor soil conditions by developing complex nutrient-conserving mechanisms, so that the loss of nutrients through drainage is compensated for by nutrients from rain and dust of the atmosphere and weathering of minerals in the soil. Furthermore, since the major part of the nutrients is usually held in biomass rather than the soil, the resulting loss through drainage water is minimal. If we consider the build-up of organic contents in the soils of such specialised ecosystems, it provides an excellent illustration of how nature maximises the advantages of tropical areas. Incidence of high temperatures, humidity and rainfall, total absence of frost and varied species diversity, contribute to a higher number of life-cycles and higher production of biomass than possible in temperate regions. As long as the systems remain closed, the nutrient cycle continues undisturbed. When the cycle is broken by the destruction of the rain forest due to farming, logging or overgrazing, the organic content of the soil is destroyed. The loss of nutrients under such conditions is extremely high. Accordingly, if the forest sites in the humid tropics are to be converted into agricultural areas, inputs of fertilizers often become necessary, since they are rapidly leached away by rain, and thus are somewhat transitory in their effects (Richards, 1977). This creates two problems: agriculture under such conditions is often uneconomic, and leached fertilizers could contribute to adverse environmental effects. Richards (1977) states that "in some areas climax forests exist under conditions of nutrient deficiency so extreme that they cannot be replaced by any form of permanent agriculture, e.g., the 'campinas' and 'pseudo-caatingas' on podzolic sands in the Rio Negro region of Amazonia and the 'kerangas' (heath) forests of Borneo."

There are other problems with tropical soil as well. Preparation of the land for planting is generally carried out prior to the onset of the rains. This means that this arduous task has to be carried out very often in what turns out to be the hottest and driest season of the year by labour-intensive means with people who are mostly undernourished. In contrast, in temperate climates, precipitation exceeds evaporation during winter months, and consequently it is comparatively easier to work with the moist soil in the spring.

BIOLOGICAL DIVERSITY

Another important area where climatic factors have a significant bearing on the development process is biological diversity. Ecosystems are formed by the interaction between abiotic and biotic environments. Climate influences both biotic and abiotic environments. Since ecosystems are fundamental building blocks for development, climate has a major bearing on the development process.

There are two principal methods for quantified classification of ecosystems at the climatic level (Tosi, 1980) — coincidentally both of which were initially postulated during the same year, 1948 — by Holdridge (1948, 1967) and Thornthwaite (1948). Of the two methods, Holdridge's World Life Zone System is more ecologically oriented, since it provides a good framework for classification of various terrestrial ecosystems on a quantifiable basis. This multifactorial classification scheme provides a predictive relationship between climatic parameters and the principal features of associated vegetation. Climatic factors considered for this classification system are long-term average annual temperature, precipitation and humidity. Each life zone defines a distinctive set of possible ecosystems, known as *associations,* that are unique to the given climate. In other words, a specific *association* will not occur in more than one life zone.

Globally, approximately 125 different bioclimates can be observed under the life zone system, although the classification system does not prescribe an absolute upper limit for such numbers since some may fall outside the climatic limits on which the system is based. A significantly higher number of more common life zones can be observed in the tropics and sub-tropics where frosts do not occur. Thus, nearly 38 life zones can be observed in the tropics, and another 30 in the subtropics. In other words life zones in the tropics and subtropics account for approximately 60 per cent of the world total. In contrast, warm temperate regions contain 23 life zones and cool temperate regions another 16. Only 9 life zones can be observed in the boreal region. In spite of this difference in diversity of life zones between the tropical and temperate regions, virtually all the industrialised countries and high-yielding, agriculturally successful cases appear to be concentrated within these two temperate, mid-latitude regions (Tosi, 1975).

The diversity is even more remarkable if specific countries are considered. An extraordinary number of life zones can be noted in a country like Peru, 71 in two regions. In part this notable diversity is due to the differing landforms, orographic and climatological conditions that exist due to the Andean mountain range (Tosi, 1980). Two small countries — Costa Rica and Panama — have 12 life zones each, a clear indication of pronounced diversity of bioclimates within a limited area. In comparison, the situation in temperate climates is very different. The Netherlands has only one life zone, and the vast geographical area of the United States, east of the 102° meridian has only 10.

As to be expected under this situation, the biological diversity of the tropics is significantly greater than in the temperate regions. In spite of this difference, the unfortunate point is that the existing state of knowledge of species in the tropical regions is highly limited. During the past three centuries, attempts have been made to classify and catalogue organisms. Currently, approximately 1.5 million varieties have been named, out of which only about one-third is from the tropics (NRC, 1982). It was noted during the 1975 Congress of the International Botanical Congress in Leningrad that while the tropics hold 70 to 90 per cent of all plant species, nearly 90 per cent of botanical taxonomic work is done on the plants of the temperate region.

A recent report by the United States National Research Council (1982) estimates that the number of species of organisms in the tropics is approximately twice the number found in temperate climates. This estimate is based on an analysis of species of birds, mammals and butterflies, which are relatively well-documented. For these groups, nearly twice as many species can be observed in the tropics as compared to temperate regions. The same NRC report provides a very rough estimate of the total number of species of plants, animals and organisms in the tropics as approximately 3 million and 1.5 million for temperate regions. This means only about 17 to 18 per cent of tropical organisms have received any scientific attention thus far.

The lack of scientific knowledge of tropical species is further highlighted in another report of the United States National Research Council (1980). It points out that information on even economically important tropical species, e.g. freshwater fishes or higher plants, is still very limited. For example, if Latin America is considered, it contains about 80,000 land plant species, which is approximately 33 per cent of all the different varieties in the world. Nearly one in eight of these plants has not yet been classified since they are still not known to scientists. If freshwater fishes in South America are considered, nearly 40 per cent of an estimated 80,000 species have yet to be discovered (Böhlke *et al.,* 1978).

The situation, as to be expected, is significantly worse for lesser known organisms. Thus, fewer than 15,000 of several hundred thousand nematodes to be found globally have been catalogued (NRC, 1982). This does not mean that they are not economically important, since many are parasites of economically useful plants and animals. Similarly, if fungi are considered, currently it is "virtually impossible to prepare regional catalogues for any area, because there are none in the tropics for which the fungi are relatively well known" (NRC, 1982). This is despite the fact that fungi currently cause tens of billion dollars of damage every year (Day, 1977).

Biological diversity is an important factor for development. Without information and knowledge of different alternatives available, it is not possible to develop sustainable, productive systems in the tropical climates that are appropriate to specific regions and conditions. This is important, especially when it is considered that the number of species on which detailed information is available is strictly limited. It is because only some 5,000 plant species have been used historically for food and fibre, out of which only about 150 are used extensively. The situation is even more skewed if human energy requirements are considered, where only 3 species — wheat, rice and maize — provide more than 50 per cent of the energy as shown in Figure 1.1.

Even though few species of plants are used at present extensively throughout the world for food, it should not be assumed that they always existed in various places all over the world. In fact, the situation is generally the reverse. The species either originated or evolved rapidly in a limited area. The eminent Russian plant breeder and

Figure 1.1. Centres of origin and variability of cultivated plants (after Vavilov, 1935).

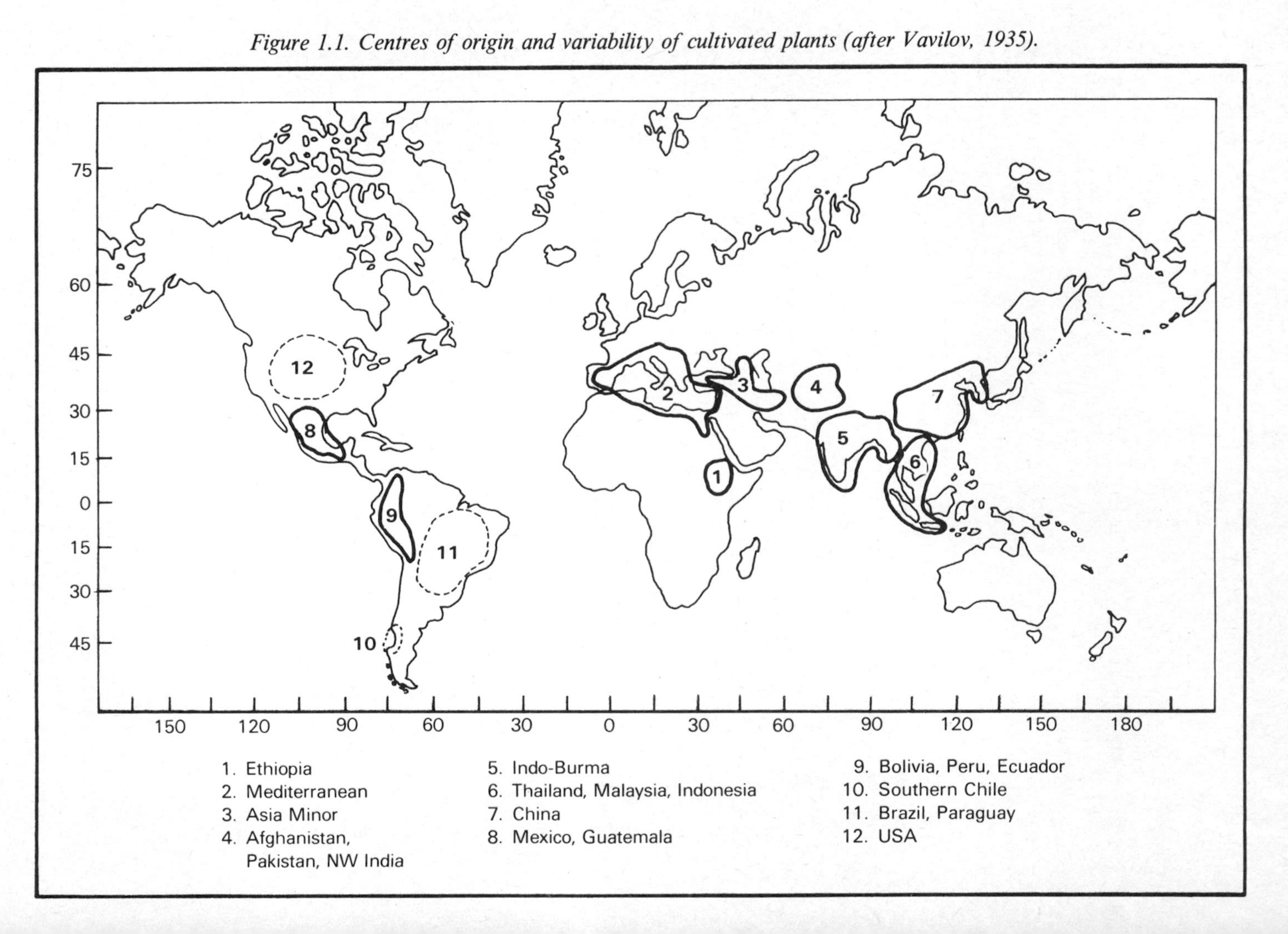

geneticist, N.I. Vavilov, analysed the areas where such development took place, and then identified 12 areas — 9 major and 3 minor — which provided much of the genetic diversity of the presently cultivated plants. These 12 centres of origin — known as Vavilov Centres — are predominantly from the tropical regions as shown in Figure 1.2.

With increasing demand for food production, the global system is becoming increasingly homogenised. The reason is relatively simple. Comparatively few species that are being used as food crops have received universal acceptance because of their taste and nutritious value, as well as being relatively easy to grow under different conditions, and readily amenable to storage, transportation and marketing.

The worldwide success of these few crops, however, could pose a serious problem for mankind in the future. The uniformity of modern agriculture means that large areas had to be cleared or are being cleared for monoculture under controlled conditions. During this clearance process, natural vegetation that existed earlier had to be destroyed. This naturally reduces the biological diversity of the area, since few species of crops are displacing the numerous varieties that existed before. Thus, crop germplasms disappear during this development process. This is especially important for the Vavilov centres of the tropics, where biological diversity is high. Accordingly, unless crop germplasms are carefully preserved, there is a real danger that these may be permanently lost to mankind. Furthermore, since our knowledge of plant species of

Figure 1.2. Human calorie sources from plants

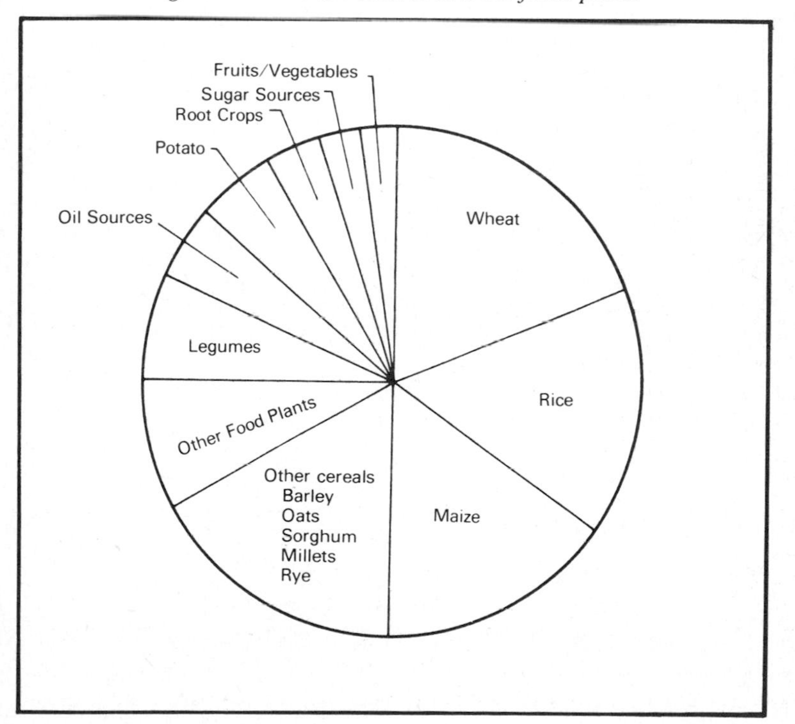

the tropics is somewhat limited, it could mean potentially useful varieties may not even be discovered. This could become a serious long-term problem.

The usefulness of biological diversity for economic development should not be underestimated. During the past few decades much progress has been made in genetic research which has contributed to the raising of the yields of major plants and animals. Because of increased yields, the general pattern of agricultural development has tended to concentrate increasingly on a few high-yielding and low-cost crops, that are resistant to certain pests and diseases. A direct consequence of this development has been greater and greater dependence on a few select varieties of crops. Thus, more than half of the Canadian prairies cultivate only a single variety of wheat — Neepawa — and only four varieties of wheat account for 75 per cent of wheat produced in that region. The situation is not much different in the United States, where 72 per cent of potato production is due to only four varieties and only two varieties account for the pea production (OECD, 1982).

A major problem that often arises from a high degree of genetic uniformity is the difficulty of controlling pest outbreaks on a massive scale in areas where monoculture is both intensive and extensive. Good crop yields for several years often tend to lull farmers into a false sense of security, and when major pest and disease outbreaks occur, the consequences could be catastrophic. Several such historical precedents exist at present.

Probably the most well-known example of such a failure is the disastrous potato crop failure in Ireland and Europe in the 1840s due to blights. Some 2 million people died due to starvation, and another 2 million emigrated (Carefoot and Sprott, 1967). Between 1870 and 1890 rust virtually destroyed the coffee production of Sri Lanka, which was then the world's largest coffee-growing nation (CEQ, 1980). The infamous Bengal famine of 1942 was precipitated by the failure of paddy production due to a fungus, which ultimately resulted in the death of thousands of people. Chestnut blight has virtually decimated American edible chestnut trees. Periodically pest outbreaks have severely reduced corn, oat and wheat production in the United States. Furthermore, severe pest outbreaks and the subsequent difficulties in controlling them can radically alter locations of monocultural plantations. Typical examples are translocation of banana plantations from the Caribbean to the Pacific side of Central America due to heavy losses from Panama wilt *(Fusarim oxysporum)*, and shifting of rubber plantations from Brazil to Asia due to massive incidence of South American leaf blight (*Microcyclus ulei*) (NRC, 1982).

As a general rule, the number of species of pests in tropical monocultural plantations increases with time. Thus, the longer a crop is grown in a specific site, the higher would be the varieties of pest (Strong, 1974). It has been suggested that a new variety of wheat can probably be cultivated for about 10 years, before its resistance is broken down by constantly evolving persistent attacks by pests and pathogens (CIMMYT, 1974).

There are some exceptions. For example, some species such as teak in India and Indonesia and *Eucalyptus* in Brazil have not had serious pest outbreaks for decades. However, the reason or reasons as to why certain species do not have serious pest problems for a prolonged period is still not known (NRC, 1982).

This general state of affairs means that farmers have to be flexible with monocultural plantations. When a particular variety of crop suffers extensively from pest outbreaks, it becomes vulnerable and may no longer be economic to grow. Better management practices can control losses on a short-term basis, but they are not long-term solutions.

Use of chemical insecticides can help initially, but climatic conditions are favourable for pest growth in the tropics. This means pests can go through many generations within a short period of time, and during this process can become immune to the pesticide being used through evolutionary adaptation. Thus, in Egypt, chemical pesticides being used for cotton pests have to be changed frequently.

Substitution of a vulnerable variety of crops with a more desirable variety is one solution, but it is not an easy process in most cases in developing countries of the tropics, where continual pest monitoring and adoption of optimal pest control practices may not be possible. In addition, lack of suitable infrastructure to give appropriate advice, especially in rural areas, and limited availability investment capital, credit and necessary agricultural inputs, severely restrict the options available.

The problem can be more serious in developing countries of tropical climates than industrialised countries of temperate zones. As mentioned earlier, biological diversity of tropical climates is much higher than in temperate climates. This holds true for plant pests and pathogens. Accordingly, crops in tropical and subtropical climates face far more varieties of pest problems than their temperate counterparts. This is clearly shown in Table 1.3 where the number of diseases reported for different crops are shown for tropical and temperate regions (Swaminathan, 1979).

In addition to a higher diversity of pests in the tropics, the climatic conditions in the tropics — temperature, precipitation, soil moisture, and sunlight — are more favourable to pests than in temperate regions. The absence of frost in the tropics means that insects, pests and parasites live and proliferate without any interruption throughout the entire year. In contrast, in temperate climates, frost and snow act as the great executioner of nature in eradicating pests.

There is an urgent necessity to develop reliable pest forecasting methods which need to be integrated with climatological factors. Such integration is likely to pay rich dividends, especially in tropical areas. Already some benefits are being obtained from the limited information available. For example, healthy seed potatoes are now being grown in Northern India, because it was found that aphids, which are vectors of several virus diseases, are absent during certain months of the year (Swaminathan, 1979).

In future, agriculture will have to depend on continued input of genetic diversity, and accordingly maintenance of biological diversity is essential. Such genetic developments have already made significant contributions to pest control and

Table 1.3. Comparison of crop diseases in tropical and temperate climates (Swaminathan, 1979)

Crop type	*Number of diseases reported* Tropics	Temperate zone
Rice	500-600	54
Corn	125	85
Citrus	248	50
Tomato	278	32
Beans	250-280	52

management. Thus, recent improvement in resistance of peanuts to leafspot has been made possible from the wild varieties *(Arachis monticola, A. batizocoi* and *A. vilosa)* of the tropical forest of Amazonia. The annual value of this development has been estimated at US$500 million (NRC, 1982).

Maintenance of biological diversity is not only essential for pest control but for other areas as well. A prime beneficiary is the pharmaceutical industry. It has been variously estimated that 25 per cent (NRC, 1978) to more than 40 per cent (OECD, 1982) of the prescriptions written every year in the United States contain drugs derived from organisms. According to OECD (1982), 25 per cent come from higher plants, 13 per cent from microbes and 3 per cent from animals. The commercial value of these preparations has been estimated from US$3 billion (Farnsworth and Morris, 1976) to over US$10 billion (OECD, 1982). Whatever may be the true value of these prescriptions, it is obvious that it is substantial. On a global basis, the figures are likely to be higher by several magnitudes. Currently, it is estimated that some 3,000 plant species possess anti-cancer properties, 70 per cent of which grow in humid tropical forests (Wilkes, 1981, as quoted by NRC, 1982).

In animal genetic resources the problem is not any less serious when compared to plant genetic resources. Like their plant counterparts, very little is known about genetic resources of domesticated tropical animal varieties such as goats, camels, water buffaloes, llamas or alpacas (FAO, 1978). Information on native cattle, poultry and pigs is also limited, and many of the indigenous breeds are threatened with extinction. The situation is also serious in many temperate regions. For example, 115 of the 145 of the cattle strains that are indigenous to Europe and the Mediterranean region are threatened with extinction.

Increasing population pressure is threatening the biological diversity of the tropics. Deforestation is a major contributory problem. At least 50 per cent of deforestation in the humid tropics can be accounted for by shifting cultivation, which currently provides marginal subsistence for some 200 million people (NRC, 1982). This practice is contributing ultimately to endangering and disappearance of many plant and animal species about which we know very little, but which could be potentially beneficial in the future. Unless germplasms are carefully preserved, it could mean substantial losses in the future.

TROPICS AND DEVELOPMENT

In the preceding discussion, an attempt has been made to show that climate is an important factor that has a direct bearing on both the biotic and abiotic community. Since economic development is clearly dependent on the available biotic and abiotic resources, climate is clearly a factor that has to be considered explicitly in any economic development theory of the tropics and the subtropics: otherwise the theory is unlikely to be viable on a long-term basis. Furthermore, the physical, social, economic and cultural conditions and institutional infrastructures are often very different in the tropics, when compared to their counterparts in temperate climates.

Table 1.4. Difference between temperature and tropical agriculture (NRC, 1982)

Conditions and practices	Temperate zone agriculture	Humid tropical agriculture
Controlling factors	Mostly physical	Mostly biological
Growing season	3 to 8 months	12 months
Dieback, frost, aridity	Common	None
Deforestation (land clearing)	Customary	Partial
Bare soil present	Common	Ideally none
Changes in water relationships	Common	Common
Nutrient cycle	Open	Partly open
Annuals and perennials	Quick annuals (3 months)	Perennials and annuals (more than 5 months)
Dominant crops	Seed	Vegetative, root, and seed
Year-to-year fluctuation in production	Wide variance	Little variance
Labour factor in productivity	Machine-intensive	Hand-labour-intensive
Planting density	High	Low
Field structure	Monoculture	Polyculture
Diversity of genotypes	Low	High
Competitors	Few	Many
Storage of products	Long-term	Short-term (fungi and pests abundant)
Individual biomass	Low	High
Food chains	Short	Long, complex
Cropping pattern	No stratification of fields	Multicropping

Thus, a development plan or process that has proved to be a success in a temperate zone may not necessarily be successful in tropical climates. For an important human activity such as agriculture, there are some fundamental differences in conditions between temperate and tropical zones as shown in Table 1.4 (NRC, 1982).

It is interesting to note how scientific opinion on the potential of tropical agriculture has changed during the past century. Discussing the agricultural developments of the late nineteenth and early twentieth centuries, Tempany and Grist (1958) have noted: "Many mistakes have been made because European planters and administrators were unaware that, under tropical conditions, factors require to be taken into account which are operative only to a minor degree in temperature countries, often coupled with an erroneous belief that tropical soils are well-nigh of inexhaustible fertility." By the second and third decade of the twentieth century, it was being gradually recognized that luxuriant tropical vegetation did not necessarily represent abundant fertility.

The optimism of the earlier decades was somehow replaced by the pessimism of many people during the sixties and seventies. The failure of many development plans convinced some people that agricultural development in the tropics would have to be limited.

This overtly pessimistic view is as inappropriate as the earlier highly optimistic expectations. It is being argued here that failures of many of the past policies and plans are to be expected because some of the fundamental assumptions on which they were based were incorrect. A reorientation of attitudes to and perceptions of development in the tropics is necessary.

Over the last two to three decades, many myths have grown about the tropics, the majority of which are pessimistic. It is often now believed that tropical soils are very low in fertility, climatic conditions are extreme, plant and animal pest problems are insurmountable, human health problems are serious, etc. While there is an element of truth to many of these issues, which presumably is the reason why these myths grew, it is neither possible nor scientific to make such generalized statements for a vast area containing very different physical, social and economic conditions.

One is tempted to consider what might have been the general feeling, if the situation was the reverse of what exists at present. Let us assume that it was the tropical countries that were developed rather than the temperate zone countries and that most of the research of the past two centuries was carried out on tropical conditions. Under this situation, the future prospect of the "under-developed" temperate regions, on which limited information was available, may not have appeared to be bright. One can easily think of the problems of the temperate zone that might have appeared to people to be almost insurmountable:

— the growing season is short for agricultural development — only between three and a half to six months. Cropping intensity cannot approach tropical conditions, and hence one crop a year has to last a family the entire year.
— Biological diversity is low. Many crops will be difficult to grow economically.
— Long winter months with heavy snowfall means high energy costs for human survival. Clothing costs will be far higher than tropical climates.
— Transportation will be difficult in winter. Navigation will be impossible because many rivers are frozen in winter.
— Domestic animals have to be protected carefully during the long winter months.
— Diseases of temperate climate, some known and others unknown, would create havoc with a large population.

If the table was thus reversed, one may very well have been pessimistic on the future prospects of the development of temperate regions. These would, however, have been false concerns, stemming directly from the lack of knowledge as to how to develop the region appropriately. We currently face the same problem with the tropical countries. As discussed earlier, our knowledge of tropical conditions is limited. We still do not know enough about tropical plants, many of which can be successfully exploited as sources of food, fibre, forage and fuel. We need more information on tropical soil, and ways by which sustainable agricultural yields can be obtained. Ways and means of carrying out scientific research have improved tremendously during the past two to three decades, and thus it should not take us another two centuries to ascertain sustainable development processes for the tropics.

The crucial problem is time. Current estimates indicate that the population of the tropical countries may increase by another 1.5 billion during the next two decades. Provision of adequate basic needs to this increasing population is not going to be an easy task. The most fundamental problem facing the planners and decision-makers is how to devise and implement appropriate development strategies for the tropical countries which will satisfy short-term requirements of immediate economic development but will have no long-term adverse irreversible consequences. When it is further considered that the right policies have to be devised in spite of inadequate scientific knowledge, the problem is going to be a most difficult one to resolve. Given the political will in both temperate and tropical countries, the problem can be solved. If not, the future will be bleak indeed.

REFERENCES

Biswas, Asit K. (1979) "Climate, Agriculture and Economic Development," in *Food, Climate and Man,* Edited by Margaret R. Biswas & Asit K. Biswas. John Wiley & Sons, New York, 1979, pp. 237-259.

Biswas, Margaret R., and Biswas, Asit K. (1981) "Environment and Development," in *Impact of Development of Science and Technology on Environment,* Edited by A. K. Sharma and A. Sharma. Indian Science Congress Association, Calcutta, pp. 107-114.

Boulding, Kenneth E. (1970) "Increasing the Supply of Black Economists: Is Economics Culture-Bound?" *American Economic Review,* Vol. 60, No. 2, pp. 406-411.

Böhlke, J. E., Weitzman, S. H., and Menezes, N. A. (1978) "Estado Atual da Sistematica dos Peixes de Agua Doce do America do Sul," *Acta Amazonica,* Vol. 8, No. 4, pp. 657-677.

Carefoot, G. L., and Sprott, E. R. (1967) *Famine on the Wind: Man's Battle Against Plant Disease.* Rand McNally, Chicago, p. 81.

CIMMYT *Review* (1974) Centro Internacional de Mejoramiento de Maiz y Trigo, Mexico, p. 7.

Council on Environmental Quality (CEQ) (1980) "The Global 2000: Report to the President of the United States," *Study Director G. O. Barney,* Vol. II, Pergamon Press, Oxford, pp. 288-292.

Day, P. R., Editor (1977) *The Genetic Basis of Epidemics in Agriculture.* New York Academy of Sciences, New York.

Farnsworth, N. R., and Morris, R. W. (1976) "Higher Plants: The Sleeping Giants of Drug Development," *American Journal of Pharmacy,* Vol. 148, pp. 46-52.

Fisher, C. A. (1961) *South-East Asia, A Social, Economic and Political Geography.* Methuen, London.

Food and Agricultural Organization (FAO) (1978) *The State of Food and Agriculture, 1977.* FAO, Rome, pp. 3-40-3-43.

Galbraith, J. K. (1951) "Conditions for Economic Change in Underdeveloped Countries," *American Journal of Farm Economics,* Vol. 33, p. 693.

Holdridge, L. R. (1967) *Life Zone Ecology,* Revised Edition. Tropical Science Center, San José, Costa Rica, 206 pp.

Holdridge, L. R. (1948) "Determination of World Plant Formations from Simple Climatic Data," *Science,* Vol. 105, No. 2727, pp. 367-368.

Hudson, G. (1971) *Soil Conservation.* Batsford, London, 320 pp.

Kamarck, A. M. (1976) *The Tropics and Economic Development.* Johns Hopkins Press, Baltimore.

Lee, D. H. K. (1957) *Climate and Economic Development in the Tropics.* Harper, New York.

Lewis, W. A. (1955) *Theory of Economic Growth.* Allen and Unwin, London, pp. 53, 416.

Misra, R. P. (1981) "The Changing Perception of Development Problems," in *Changing Perception of Development Problems.* Edited by R. P. Misra and M. Honjo, Maruzen Asia, Singapore, pp.7-37.

Morse, B. (1980) *United Nations Development Programme in 1979: Report and Review.* UNDP, New York, 24 pp.

Myrdal, G. (1968) *Asian Drama: An Inquiry into the Poverty of Nations.* Vol. III, Pantheon, New York, pp. 2121-2138.

National Research Council (NRC), Committee on Selected Biological Problems in the Humid Tropics (1982) *Ecological Aspects of Development in the Humid Tropics.* National Academy Press, Washington, D.C., 297 pp.

National Research Council (NRC), (1978) *Conservation of Germplasm Resources: An Imperative,* National Academy of Sciences, Washington, D.C., 116 pp.

National Research Council (NRC) (1980) *Research Priorities in Tropical Biology,* National Academy of Sciences, Washington, D.C., 116 pp.

Organisation for Economic Co-operation and Development (OECD) (1982) *Economic and Ecological Interdependence.* OECD. Paris, pp. 35-37.

Ormerod, W. E. (1978) "The Relationship Between Economic Development and Ecological Degradation: How Degradation Has Occurred in West Africa and How Its Progress Might Be Halted," *Journal of Arid Environments,* Vol. 1, pp. 357-379.

Schnitzer, M (1976) "The Chemistry of Humic Substances," in *Environmental Biogeochemistry.* Edited by J. O. Nriagu, Ann Arbor Science Publications, Ann Arbor, 426 pp.

Stamp, L. D. (1966) *Asia: A Regional and Economic Geography.* Twelfth edition, Methuen, London.

Strong, D. R. (1974) "Rapid Asymptotic Species Accumulation in Phytophagous Insect Communities: The Pest for Cacao," *Science,* Vol. 185, No. 4156, pp. 1064-1066.

Swaminathan, M. S. (1979) "Global Aspects of Food Production," *Proceedings, World Climate Conference.* World Meteorological Organization, Geneva, pp. 369-406.

Tempany, H., and Grist, D. H. (1958) *An Introduction to Tropical Agriculture.* (Reprinted with corrections, 1960), Longmans, London.

Thornthwaite, C. W. (1948) "An Approach Towards a Rational Classification of Climate," *Geographical Review,* Vol. 38, pp. 55-94.

Tosi, J. (1975) "Some Relationships of Climate to Economic Development in the Tropics," in *The Use of Ecological Guidelines for Development in the American Tropics.* New Series No. 31, International Union for Conservation of Nature, Morges, Switzerland, pp. 41-55.

Tosi, J. (1980) "Life Zones, Land Use and Forest Vegetation in the Tropical and Subtropical Regions," in *The Role of Tropical Forests on the World Carbon Cycle.* Edited by S. Brown, A. E. Lugo and B. Liegel, Center for Wetlands, University of Florida, Gainesville, pp. 44-64.

Uehara, G. (1981) "Tropical Soil Management and Its Transfer," *Food and Climate Review 1980-1981.* Edited by L. E. Slater, Aspen Institute for Humanistic Studies, Boulder, pp. 47-51.

United Nations Industrial Development Organisation (UNIDO) (1980) "Report of the Third General Conference, New Delhi, India," *Report ID/CONF. 4/22,* UNIDO, Vienna, 143 pp.

Wilkes, H. G. (1981) "New and Potential Crops, Or What to Anticipate for the Future," Paper Presented at Annual Meeting, American Association for Advancement of Science, Toronto.

World Bank (1982) *World Development Report 1982.* Oxford University Press, New York, 172 pp.

Young, A. (1976) *Tropical Soils and Soil Survey,* Cambridge University Press, Cambridge, 330 pp.

CHAPTER TWO

Climate and Health

H. E. Landsberg
Department of Meteorology, University of Maryland

THE NOTION THAT climate is related to health dates back to antiquity. In the famous health compendium of Hippocrates (400 to 370 BC approximately), the Aphorisms, in Section III not less than 23 of the short categorical statements refer to the effects of seasons and weather on health. Thus we read in Aphorism 15 of Section III: "of the constitutions of the year, the dry, upon the whole are more healthy than the rainy and attended with less mortality."

Climate and weather continue to play a conspicuous role in human health. They govern our comfort, they may promote disease or affect disease vectors, but they can also support healing and favour recreation. Thus there are wide geographical differences in the bioclimate affecting humans. The physiological and pathological reactions to the atmospheric environment have been in part brought about by the spreading of humans all over the globe. This has brought many into regions where they are subjected to rigours for which nature has not endowed them. Only potent engineering methods, including housing, clothing, heating, and cooling permit survival and capacity to work efficiently in many areas.

The climate, which is nothing but the composite of many individual weather situations over a longer interval of time, in most localities exhibits a notable annual course. There are cold and warm, wet and dry, windy and calm seasons. These rhythms have profound biotropic influence and often steer the course of human well-being and of illness.

Weather elements can act singly or in concert to bring about effects in our bodies. There is a long list of components which may induce reactions. This includes atmospheric composition and suspensions; solar, sky and earth radiation; air temperature; atmospheric water vapour content (humidity); wind speed; and — generally indirectly — precipitation and electrical phenomena.

Also, we cannot overlook the direct hazards of atmospheric events. Violent storms cause many casualties and deaths each year. In the United States alone, in spite of a good warning system, annual tornado deaths average over 100 and a set of such storms on April 3 and 4, 1974 caused 315 deaths and over 6,000 injuries. Lightning in the US kills, on an average, 200 persons annually and injures about 500. Sudden floods are also major killers and storm surges along coast lines where hurricanes and typhoons occur have caused mass casualties. Thus, a tropical cyclone in November 1970 in the Bay of Bengal swept over 20,000 people in Bangladesh to a watery grave, and a West Indian hurricane on September 21, 1974 killed 8,000 in Honduras.

The focus of our discussion here will, however, be on the more subtle effects of the ever-changing atmosphere on individuals and populations.

EFFECT OF SINGLE FACTORS

Some of the atmospheric elements will provoke single effects in humans. These are principally the partial pressure of oxygen and the radiation field surrounding man.

Oxygen (O_2)

Molecular oxygen (O_2) is a normal constituent of the atmosphere. At sea level it is about 21 per cent by volume of the surrounding air. With nitrogen (N_2, about 78 per cent) it constitutes 99 per cent of air. The outstanding one per cent is a mixture of minor constituent gases of which only one, radon (Rn) has biological importance, because it is a natural radioactive element emanating from the soil. Invisible water vapour is also an ubiquitous admixture to the air. We will encounter its influence later. Oxygen, the essential element in respiration, is an absolute necessity for humans and animals. It is used for oxidation in the metabolic process, which furnishes the energies used for survival. The need for oxygen increases with higher levels of metabolism, which in turn is geared to the degree of activity.

Table 2.1 shows the decrease of the partial pressure of oxygen with elevation for the lowest levels in the atmosphere where appreciable populations live. Mountain climbers in the high Andes or Himalayas will encounter even lower values.

The table also indicates the approximate saturation of the oxygen-carrying constituent of blood, the haemoglobin (Hb). For the person adapted to sea level pressure a rapid change to higher levels will lead to hypoxia with accompanying symptoms of distress, including rapid breathing, giddiness and tachycardia. This may eventually lead to complete collapse. Airplanes flying above the 2 kilometre level therefore require pressurisation.

In the lowest few kilometres healthy persons can gradually acclimatise to lower oxygen pressure. Full acclimatisation may take several months. During that time lung

Table 2.1. Partial pressure of oxygen as function of elevation

Elevation m	0	1,000	2,000	3,000	4,000
p_{O_2} millibar (Hectopascals)	204	188	167	139	123
Saturation of haemoglobin	100	97	94	90	80

capacity increases due to an increased number of alveoli and the number of oxygen-carrying red blood corpuscles (erythrocites) multiplies. Discomfort will also occur in persons adapted to life in higher elevations by a rapid move to sea level. Again several months of acclimatisation are needed to restore physiological equilibrium.

Radiative factors — Ultraviolet

Humans are immersed in radiant energy from space and the environment. They also exchange radiative energy with that environment. The electromagnetic spectrum of radiation extends over 24 orders of magnitude in the frequency domain. Many portions have biological effects. Among these are x-rays (10^{16} to 10^{20} Hz); γ-rays (10^{20} to 10^{22} Hz); cosmic rays (10^{22} to 10^{24} Hz); radar waves (10^{8} to 10^{10} Hz); microwaves (10^{10} to 10^{12} Hz). But most important in a climatic sense is solar radiation, which ranges from 10^{12} to 10^{16} Hertz.

In bioclimatic practice the subdivisions are usually expressed in wavelength rather than frequencies. Three spectral regions are distinguished: The visible (400 to 800 nm), the ultraviolet (100 to 400 nm) and the infrared (800 nm to 1 μm). The radiative flux from the sun in travelling through the atmosphere is considerably modified by absorption and scattering. The shortest wavelengths of ultraviolet radiation, which constitutes about 7 per cent of the solar radiation arriving at the earth, are most weakened. Their nemesis is the ozone layer (O_3) which has its maximum in the stratosphere at about 25 kilometres. This layer, formed by photochemical reactions, is of greatest biological importance. It will not permit any appreciable amounts of the short-wave ultraviolet (UV-C), <280 nm, to penetrate to the surface.

The medium-wave ultraviolet (UV-B), 280 to 315 nm, and the long-wave ultraviolet, (UV-A), 315 to 400 nm are weakened also in the atmosphere but will reach the surface both as direct and as scattered flux. Their intensity depends greatly on suspended atmospheric particles. This intensity is largely dependent on solar elevation above the horizon. Thus midday and, in higher latitudes, summer, are the times when ultraviolet radiation reaches its peak.

Ultraviolet solar radiation has a number of biological effects. The shorter wavelengths (UV-C, UV-B) will activate ergosterol to produce vitamin D in the skin and thus will be beneficial by preventing rickets and spasmophilia. These rays are also germicidal. On the other hand lengthy exposure will cause sunburn (erythema). This is not only painful but can lead eventually to skin cancer. The risk of skin cancer, mostly basal cell or squamous carcinoma, is greater in certain ethnic groups than others. It is quite hazardous for Caucasians of Celtic extraction. In contrast, highly pigmented ethnic groups will not suffer from ultraviolet-induced skin cancer (Belisario, 1959). Studies in Hawaii with a multi-racial population have shown that skin cancer rates in Caucasians are 138 per 100,000 population and only 1.6 per 100,000 in native Hawaiians, with a 3.1 rate per 100,000 for all non-Caucasians (Quisenberry, 1962). It must be noted also that there are other causes of skin cancer than ultraviolet radiation. Early detected basal cell or squamous skin carcincoma is readily cured. This is not the case for malignant melanoma. There is a strong suggestion that this form of cancer can also be activated by ultraviolet radiation. Figure 2.1 shows US cases of melanoma as

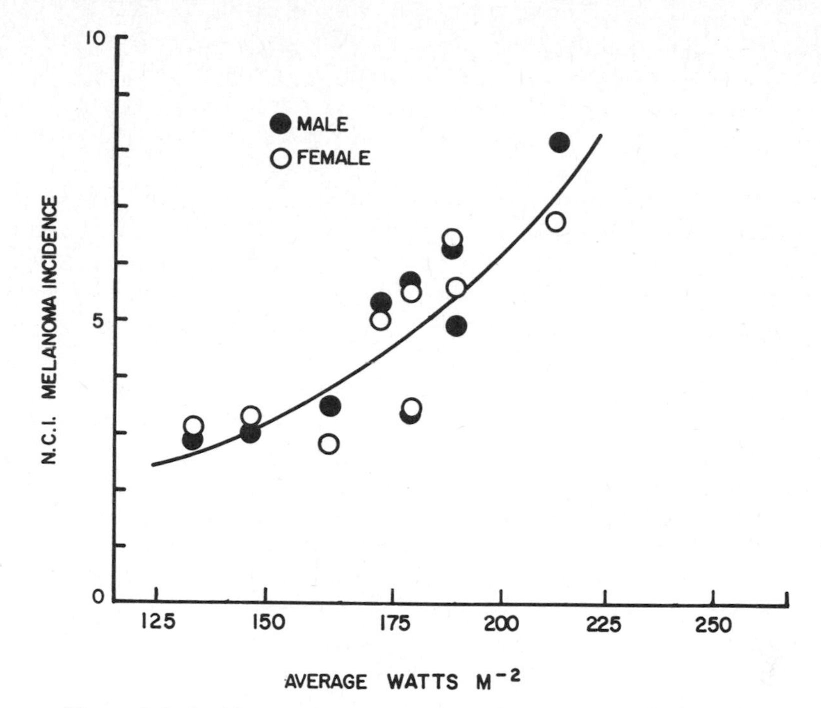

Figure 2.1. Incidence of malignant melanoma in the US as related to total incident solar (global) radiation (NCI = National Cancer Institute)

related to total average daily solar radiation. There is strong correlation between the ultraviolet portion of the radiative flux and the total. The rise of melanoma cases with the incident radiation is certainly indicative of a relation.

The etiology of skin cancer development is assumed to occur via the destruction of DNA. It is still controversial whether there is natural repair of the damage for small doses. The first sign is usually keratoses or thickening of the skin. And it is quite clear from the skin cancer incidence that there is likely to be cumulative effects of ultraviolet exposure. The age-related statistics show that in the United States there are about 2 cases per 100,000 people in the 20-year age group, 20 cases per 100,000 in 50-year olds, and about 200 per 100,000 for 70-year old persons. The fact that neck and facial features are the most common locations of skin cancer indicates solar influence. Persons in outdoor occupations are also most affected (Haenszel, 1962; Segi, 1962).

Ultraviolet A also has biological effects. It activates melanin in the skin and thus causes tanning. Persons who get sunburn and have difficulty developing a tan are most at risk of skin cancer. In this context it has been suggested that human skin pigmentation was genetically developed as protection against ultraviolet radiation and was lost by ethnic groups living in predominantly cloudy regions (Loomis, 1967).

Possible anthropogenic effects on the ozone layer have been widely cited as threatening increases in the skin cancer rates in the future (National Academy of

Sciences, 1979). Although the formidable array of stratospheric chemistry is not yet completely understood, man-made chlorine compounds, such as chlorofluoromethanes have been cited as agents causing ozone reduction. For every one per cent decrease in ozone a two per cent increase in skin cancer incidence has been projected (National Academy of Sciences, 1975; Ambach, 1978).

The ultraviolet radiation intensity, as do all solar radiation fluxes, depends on solar elevation above the horizon. Thus the greatest dosages are received close to solar noon. In higher latitudes the summer values are considerably higher than those in winter. Atmospheric turbidity plays an important role. In urban areas with high values of air pollution by particulates the solar ultraviolet is notably weakened. Another factor is cloud cover and type, and finally sea-level elevation. Table 2.2 shows the average weakening of solar ultraviolet radiation in cloudless sky depending on solar elevation. The rapid drop-off is notable. The values of the daily totals between summer and winter in middle latitudes on clear days show an order of magnitude difference.

There is also a substantial increase in solar ultraviolet with altitude. Measurements resulted in the relative values shown in Table 2.3 (Robinson, 1966).

Total radiation on a horizontal surface increases in the lower atmospheric layers by 10 to 12 per cent for each 1,000 metres altitude difference, but the total ultraviolet (all bands) increases by about 15 per cent per 1,000 metres. Actually the ultraviolet load can be notably increased by various reflective surfaces. Among them are bright cumulus clouds in the sky quadrant away from the sun. Also, highly reflective surfaces will increase the exposure, such as a fresh snowfall which may reflect 80 per cent of the incoming short-wave solar radiation or a white sand, which may reflect 30 per cent. This extra radiation explains why skiers in the high mountains even in winter and sun bathers on sunny beaches get rapidly tanned or acquire sunburns.

Radiative factors — visible and infrared

Most of the radiative exposure of humans is to visible and infrared portions of the spectrum. The visible, aside from energy exchange, affects the eye sight. This part of

Table 2.2. Relative intensities of solar ultraviolet (in per cent) in dependence of solar elevation

Solar elevation, degrees	90	70	50	40	30	20	10
Increase in optical air mass %	0	6	30	55	100	190	460
Decrease in ultraviolet %	100	86	57	41	23	11	4

Table 2.3. Ultraviolet-B radiation as function of altitude, in per cent of low-level value (after Robinson, 1966)

Altitude m	200	500	1,000	1,500	2,000	2,500	3,000	3,500
Summer	100	125	145	170	182	190	195	200
Winter	100	150	220	280	330	390	440	480

the spectrum is highly dependent on cloudiness. On overcast days at sea level in summer only one-quarter of the illumination may reach us compared to clear days. In winter, depending on latitude, the ratio may be even more unfavourable. In individual hours with thick cloud cover, there is even in midday, less than 3 per cent of the illumination than on a day with sunlight and bright cumulus clouds in the sky. In winter thick cloud cover will generally require artificial light for adequate illumination. Light has effects on the vegetative nervous system and in humans as in animals can stimulate certain hormonal secretions. Changes in illumination can also provoke psychological reactions. Lack of light can be the cause of depressions.

Specific effects of bright light and dazzling have been found in many (but not all) persons suffering from migraine. Such conditions can be often encountered in the presence of snow cover. In sensitive persons, by as yet unknown phototropic chemical reactions, vasoconstriction, leading to these headaches, are provoked. Avoidance of intense illumination and wearing of dark glasses often help (Tromp and Faust, 1977).

Infrared radiation has no known damaging effects but may occasionally provoke what has been described as heat erythema. This is better labelled as hyperemia by dilation of peripheral blood vessels which greatly increases the blood flow to the skin.

The main effect of the combined solar radiation fluxes is their influence on the human energy balance. They are an essential part of the complex system of metabolic heat loss from the body to the environment and the heat gain from the environment. These must remain in equilibrium for survival. The radiative interplay is quite complex and comprises a number of fluxes. The radiative income is composed of direct radiation from the sun, solar radiation scattered by molecules and particulates (sky radiation), radiation reflected from clouds, solar radiation reflected from the ground (as for example, by sand and snow), long-wave radiation from the ground and from radiating gases in the atmosphere, such as water vapour and carbon dioxide. In addition there is long-wave radiation from objects, such as walls or trees. The human body also loses energy by such infrared radiation to the sky, the ground, and surrounding objects. Figure 2.2 shows these various fluxes. Mathematically the net energy gain or loss by radiation can be expressed as follows:

$$(1) \qquad \pm Q_N = Q_s + Q_{sc} + Q_{rc} + Q_{LG} + Q_{LA} + Q_{LO} - Q_{rB} - Q_B$$

where Q_N = net energy gain or loss
Q_s = total energy in direct beam of the sun
Q_{sc} = short-wave energy scattered by sky and particles
Q_{rc} = short-wave energy reflected from clouds
Q_{rg} = short-wave energy reflected from the ground
Q_{LG} = long-wave energy emitted by the ground
Q_{LA} = long-wave energy back-radiated by the atmosphere
Q_{LO} = long-wave energy emitted by objects in the environment
Q_{rB} = reflected short-wave energy from body (or clothing)
Q_B = long-wave energy emitted by the body (or clothing)

The reflected energies are determined by the albedo (fraction of the incident radiation reflected from a surface) of the various surfaces. The long-wave radiation emitted by various surfaces, including the body is determined by their emissivity e and

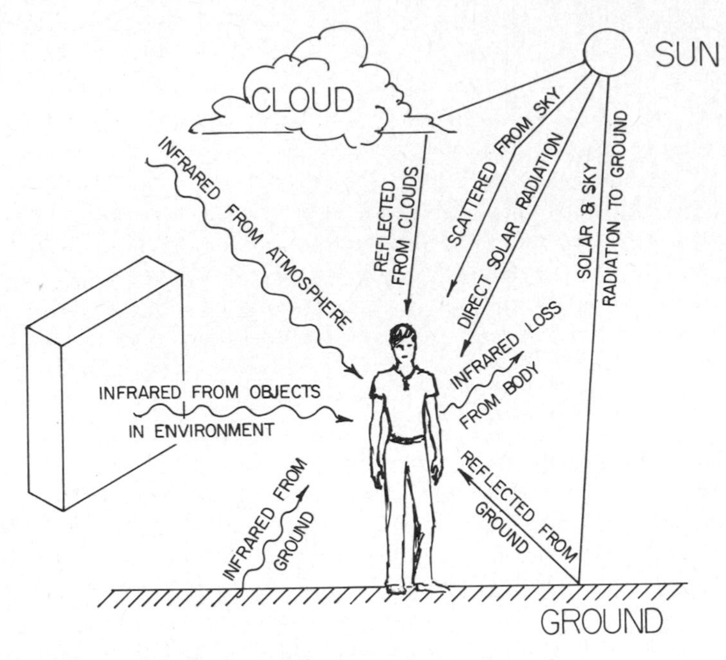

Figure 2.2. Radiative fluxes producing the radiative energy exchange between body and environment

by their absolute temperature T (in °K):

$$(2) \qquad Q_L = e\sigma T^4$$

where σ is the Stefan-Boltzmann constant 5.67×10^{-8} mW $cm^{-2}K^4$.

The albedo of human skin is only about 1 to 2 per cent in the ultraviolet and averages around 5 per cent of the incoming radiation in the infrared. In the visible there is a wide range depending on the individual wavelengths and skin pigmentation. In first approximation, light skin will reflect about 30 to 40 per cent and pigmented skin 15 to 20 per cent of the visible radiation. The energy loss of bare human skin to a clear sky, at two skin temperatures, for various air temperatures and a relative humidity of 50 per cent is shown in Table 2.4 (Buettner, 1938).

The radiative heat loss at low temperatures is very severe. One can estimate that an adult person with 1.6 m^2 body surface will lose 16 kw h per day. The daily metabolism will produce 4 kw h so that a deficit of 12 kw h has to be covered by radiative heat from the environment. Should the temperature of the radiating surface of a human at any time drop more than 7°C below that of the environment, the radiative loss will exceed

Table 2.4. Heat loss of skin against sky at various air temperatures

Air temperature °C		−10	0	10	20	30	
Skin temperature	20°C (293°K):	230	197	153	96	49	W m^{-2}
	30°C (303°K):	292	254	207	151	103	W m^{-2}

the metabolically produced heat and endanger the person. This can be the case at night under a clear sky, especially in deserts where the atmospheric counter radiation is very low due to the very low water vapour content of the air. Indoors the radiation from floors, ceiling, and walls with suitable heating devices will make up the deficit. Special clothing reflecting some body radiation back is helpful. The albedo of clothing which can reduce heatload on days with intense direct solar radiation has been determined at 60 per cent for white silk and 45 per cent for light flannel. On the other hand nylon has only 10 per cent albedo and dark wool even less than that. Other properties of clothing will be discussed later.

The distribution of solar radiation over the globe is quite complex although latitude plays a major role. It governs in particular the seasonal variation. The polar regions remain, of course, dark during their respective winter months and have a relatively high radiation income in summer. The high latitudes, just below the Arctic circle also have low amounts of radiation in winter. In the southern hemisphere there is also no land and no settlements in those latitudes but in the northern hemisphere vast land areas with considerable population exist there. The meteorological conditions, especially in northern and northwestern Europe, also cause much cloudiness so that the cold season is distinctly gloomy. This has been invoked by some psychologists as a contributing factor to alcoholism and high suicide frequency in some countries in that area. In low latitudes the season variation is not great but there are considerable differences between the low-latitude deserts and rainforests. These deserts have the highest direct radiative energy income but their high albedo reflects much back to space. The combination of these two short-wave fluxes place considerable burdens on humans in daytime. A general conception of the important direct solar and sky flux as it is received on a horizontal surface can be gained from Figure 2.3. An attempt to evaluate the various fluxes as they affect man, by latitude and elevation has been made by Terjung and Louie (1971) to whose formulations and extensive tables we refer the reader.

COMPLEX FACTORS AFFECTING THE THERMAL BALANCE

Metabolic equilibrium

Radiation fluxes, albeit very important, are not alone in affecting the thermal balance of humans. Air temperature, air humidity and air motion usually in combination play a major role. For homeotherm living beings, which includes *Homo sapiens*, maintenance of thermal equilibrium is absolutely essential for survival. We can express this formally:

$$M \pm Q_N \pm C - LE = 0 \quad (3)$$

where M = heat generated by metabolism
Q_N = net radiation
C = heat gained or lost by convection
LE = heat lost by evaporation

Figure 2.3. Average annual radiation flux from sun and sky to a horizontal surface over the globe in KWh $m^{-2} yr^{-1}$

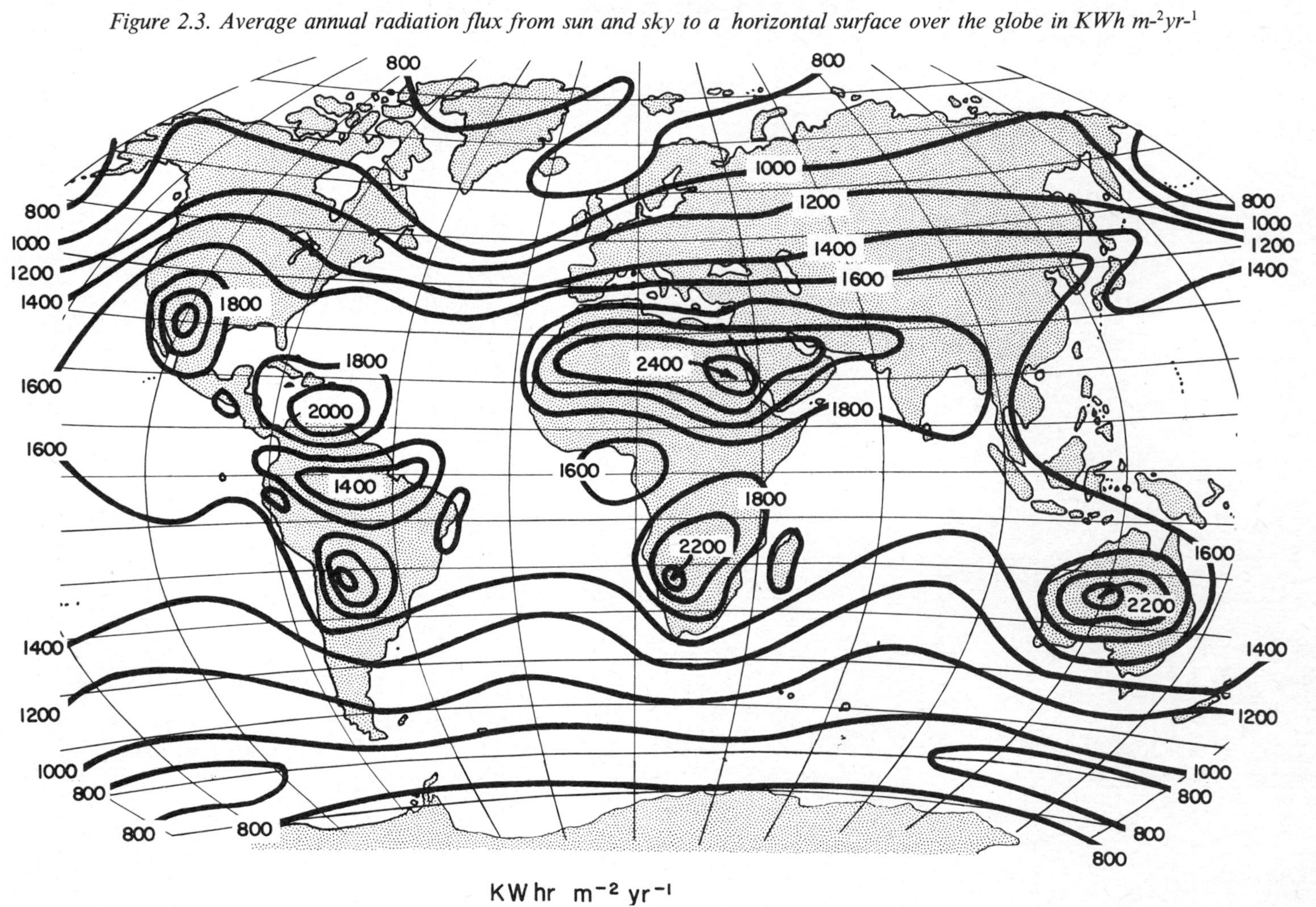

The quantity M is governed by a person's level of activity. The physiologists have conventionally given this in metabolic units or METs. One MET is defined as body heat production of 50 kilogram calories per square metre of body surface (581.5 W h m^{-2}).[1] One MET represents the metabolism of an non-sleeping person at rest. This is the so-called basal metabolic rate. The relative values of this unit are shown in Table 2.5 for various levels of activities.

The convective factor involved in the thermal equilibrium condition is itself composed of the temperature difference between body (or body surface), the surrounding air and the ambient wind speed. The evaporative factor also has two components: heat lost from evaporation of perspiration from the skin and heat lost by expiration of warm saturated air from the lungs. In the symbol of formula (3) L is the latent heat of vaporisation from water to vapour and E the amount of water evaporated. At 35°C the value of this quantity is about 6.7 kw h $m^{-2}g^{-1}$. The heat loss by respiration is mostly from evaporation but a small amount of heat is lost by warming the cool inhaled air. Table 2.6 shows the usual partitioning of heat loss.

Table 2.5. Metabolic rates in humans during various activities

Activity	Number of MET units
Sleeping	0.8
Resting awake	1.0
Standing	1.5
Light office work, driving	1.6
Standing, light work	2.0
Moderate work; walking on level ground 4 km hr^{-1}	3.0
Moderately hard work; walking 5.5 km hr^{-1}	4.0
Sustained hard work; walking 5.5 km hr^{-1} with 20 kg pack	6.0
Very heavy activity; mountain climbing; athletic competition	10.0

Table 2.6. Shares of various heat loss mechanisms of humans

Process	Average per cent loss
Radiation	60 to 70
Evaporation	20 to 25
Convection	8 to 12
Conduction (to ground, or support)	2 to 4

1. Body surface is a function of height (l in cm) and weight (p in g). It can be calculated in m^2 from the empirical formula: $A = 71.84\ l^{0.725}\ p^{0.425}$.

Survival is predicated on maintaining a nearly constant core temperature of 37°C. Peripheral and skin temperatures of 33°C provide a gradient which will permit a heat flux to the environment to rid the body of metabolic heat. The maximum tolerable variation of core temperature is about ± 4°C. At 32°C consciousness is lost by hypothermia and at 41°C hyperpyrexia will cause collapse of the circulatory system. Death will occur for core temperatures below 28°C and above 43°C. Air temperatures beyond these limits are quite common on earth, far more at the lower end than at the higher. There is some physiological adaptability but the values of human homeothermal conditions in relation to the environmental thermal factors clearly demonstrate that humans in their natural state must have developed in tropical regions. It also follows that survival in cold regions depends on clothing, housing, and artificial heat production. It is clearly evident that modern humans have literally engineered their independence from climate. The limits to this will be separately discussed for cold and hot climates.

Humans in cold surroundings

Shelter and clothing are the first line of defence against cold. But humans are not entirely dependent on these artificial means. Short exposures to mild cold can be compensated for by physiological reactions. One of these is vasoconstriction in the extremities. This will reduce the skin temperature and hence lower the radiative heat loss. This has the concomitant result of numbing finger and toes, and in the case of fingers, will notably decrease dexterity. For example, exposing healthy young men without clothing in Norwegian laboratory tests by lowering a tolerable 28°C environmental temperature to 25°C, caused toe temperatures to drop by about 3°C and average skin temperatures to fall by about 1°C (Andersen *et al.,* 1965). A second defence mechanism is a stepped up metabolism which in many cases manifests itself as shivering, an involuntary muscular activity. In the Norwegian experiments a 15 per cent increase in resting metabolism was observed.

It may be noted here that voluntary food intake is considerably higher in cold regions than in warm regions. Available information indicates a linear increase with environmental temperature decrease. Within the limits of 10°C to −15°C sustained average environmental temperature the change in voluntary caloric intake for young men in fairly strenuous work is

$$VMI = 35\Delta T + 3600 \qquad (4)$$

where VMI = voluntary metabolic intake in calories
ΔT = temperature difference below 10°C, in°C.

Natives of cold climates and acclimatised persons usually adjust their diet to include high-calorie foods, particularly fats. It has been shown for Eskimos in Alaska that their metabolism is about 30 to 40 per cent greater than that of unadapted persons. Their peripheral blood flow is also markedly better than that of unacclimatised individuals. It might be noted that blood flow to the fingers can be increased by biofeedback exercises. yet experience has shown that environmental cold is a severe handicap for human efficiency. This has been shown principally for the use of tools

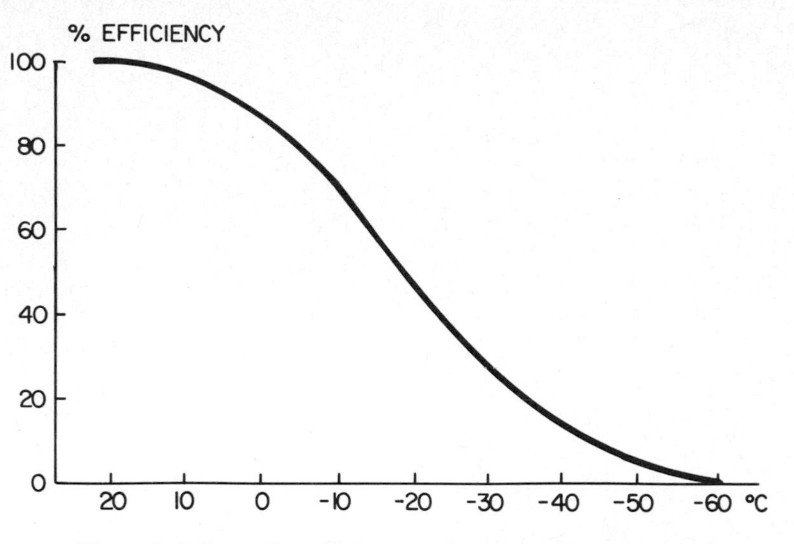

Figure 2.4. Dexterity efficiency as function of environmental temperature

outdoors. Figure 2.4 shows this for temperatures between 18°C, where dexterity is 100 per cent, and –60°C where it is 0 per cent and where all effort is essentially devoted to survival.

Actually temperature alone is an inadequate measure of environmental cold. It is rather a combination of temperature and wind speed which governs the thermal sensation outdoors in cold climates. This combination has been termed wind chill. The wind's role is the increase in convective heat loss it produces. A wind chill index was first developed during Antarctic expeditions. It was based on the amount of time it took, under given environmental conditions, to freeze a particular quantity of water. Translated into calories Siple and Passel developed an empirical model of atmospheric cooling power of the following form.

$$C = (10\sqrt{u} + 10.45 - u)(33 - T_a) \qquad (5)$$

where C = cooling power in kcal m^{-2} h^{-1}
u = wind speed, m sec^{-1}
T_a = air temperature, °C.

Cooling values obtained from this formula are represented by a series of lines for different combinations of wind speed and air temperature in Figure 2.5.[2]

For lay persons the wind chill index is not a readily understood concept. It has therefore also been converted into a wind chill equivalent temperature, also called chill factor. These equivalent temperatures are shown (both in English and Metric units) in Table 2.9. These temperatures depict the equivalent sensation felt for a calm air at that temperature compared with the prevalent temperature and wind. In full sunshine these

2. Note that the conversion factor for kcal to watt hours is 1.163, for use in the SI.

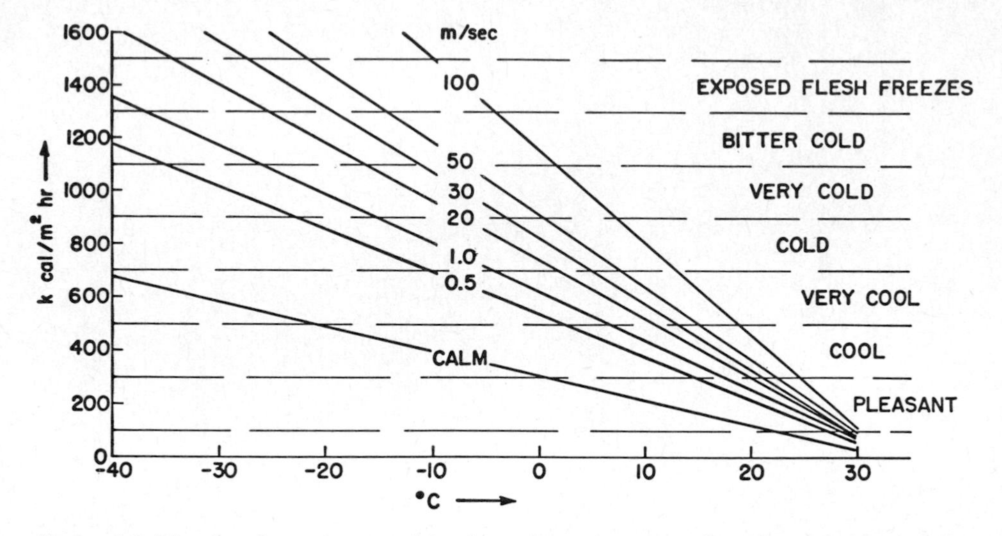

Figure 2.5. Heat loss by various combinations of temperature and wind speeds (wind chill)

wind chill equivalent temperatures are modified by the radiation received. Steadman (1971) estimated that the increase is about 14° in calm air and about 7° in strong wind. It might also be noted here that for a person walking at a 5 km h^{-1} pace in a cold environment, the wind chill is calculated for the ambient temperature and a wind speed of 1.4 m sec^{-1}.

Gregorczuk (1971) has calculated average wind chill values for the globe. The distribution of these values is shown for January and July in Figures 2.6 and 2.7. In the respective winter hemisphere the values range from > 1400 kcal m^{-2} h^{-1} (freezing of exposed human flesh) in polar regions to < 200 kcal m^{-2} hr^{-1} (pleasantly warm). In the summer hemisphere the ranges are in the north < 50 > 800 kcal m^{-2} h^{-1}, in the south < 100 to > 1200 kcal m^{-2} h^{-1}. These values only give a very crude first indication of the geographical distribution of bioclimate and it must be remembered that there are large standard deviations attached to these values.[3] From work by Flach and Morikofer (1962-67) one can gather that for mean winter-month values in the Arctic the standard deviation is about 150 kcal m^{-2} h^{-1} and in the Mid Atlantic States of the US about 100 kcal m^{-2} h^{-1}. The average diurnal variation is also fairly large. It is shown for three locations in Figure 2.8. That figure also shows another frequently used sensation scale, the cooling power which can be measured by various devices such as a katathermometer or a frigorimeter. These measurements can be approximated by the following formula

$$(6) \qquad CP - (0.23+0.47\, u^{0.52})(36.5-T_A)[\text{mcal cm}^{-2}\ \text{sec}^{-1}]$$

where u = wind speed in m sec^{-1}
T_A = air temperature, °C.

3. More detailed charts are available for several countries: US (Bristow, 1955; Terjung, 1966); Poland (Gregorczuk, 1971); UK and Ireland (Mumford, 1979).

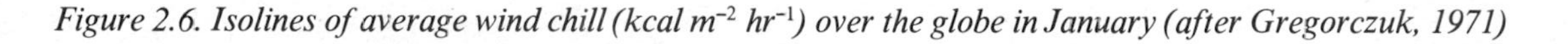

Figure 2.6. Isolines of average wind chill (kcal m^{-2} hr^{-1}) over the globe in January (after Gregorczuk, 1971)

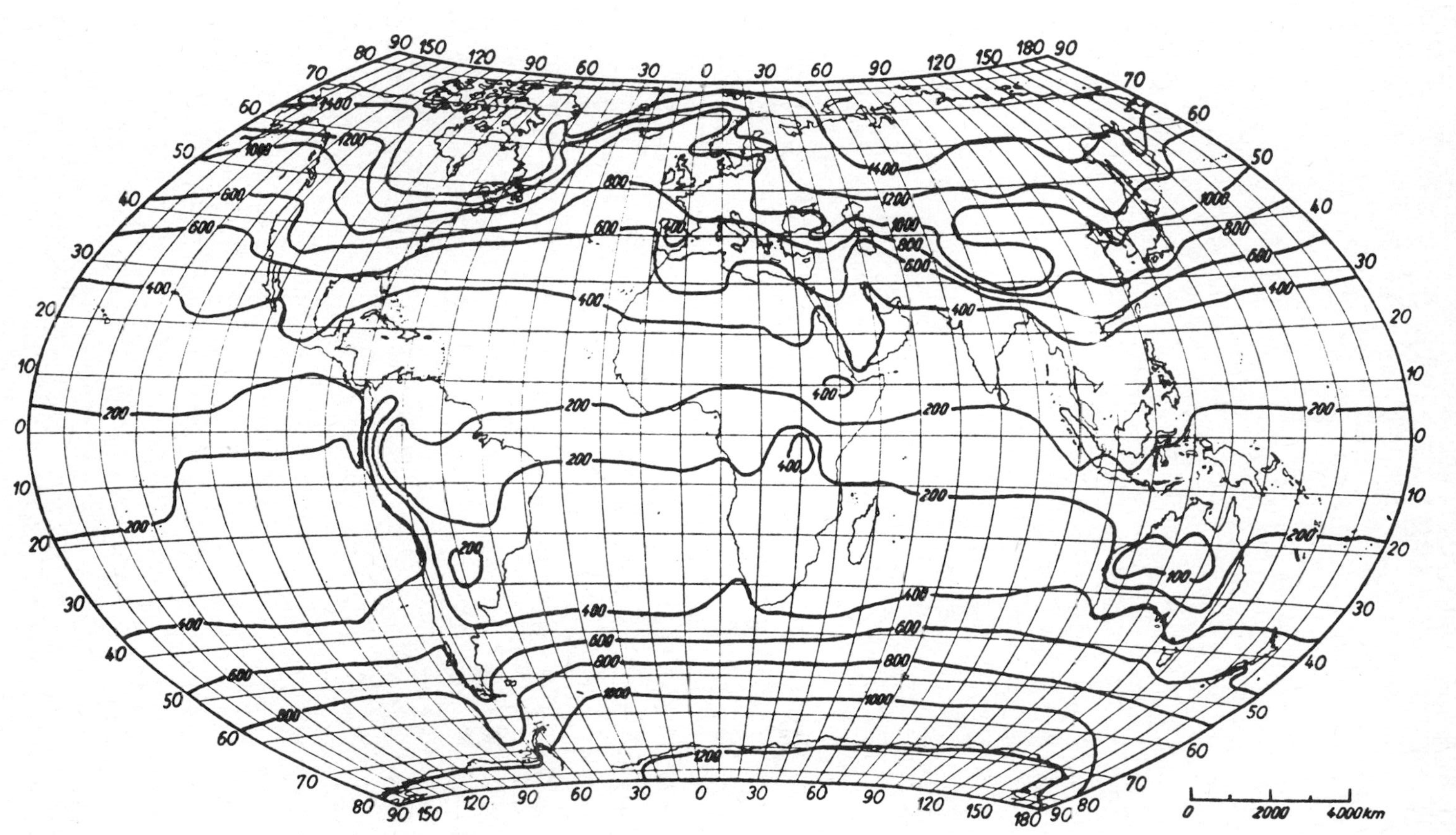

Figure 2.7. Isolines of average wind chill (kcal m^{-2} h^{-1} over the globe in July

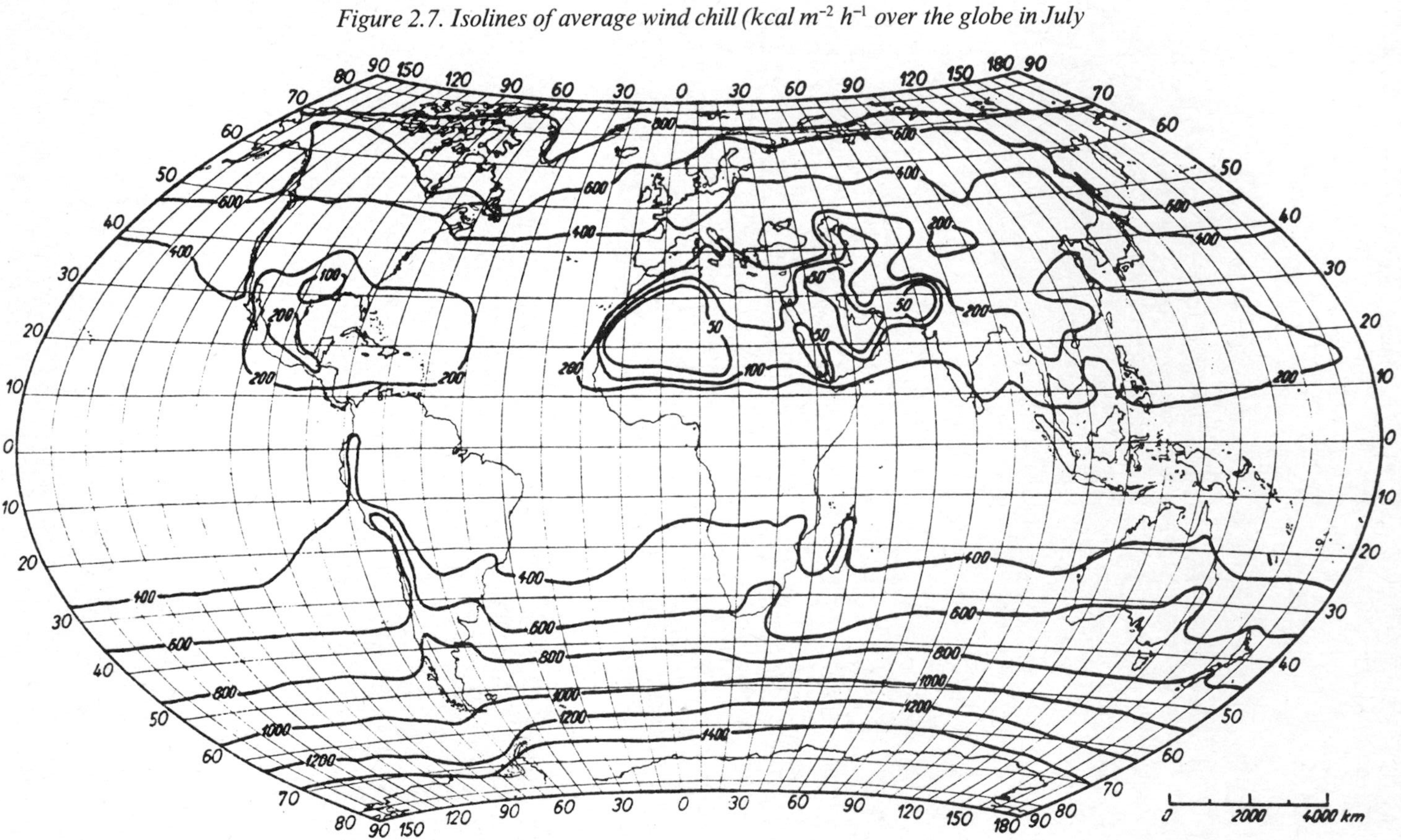

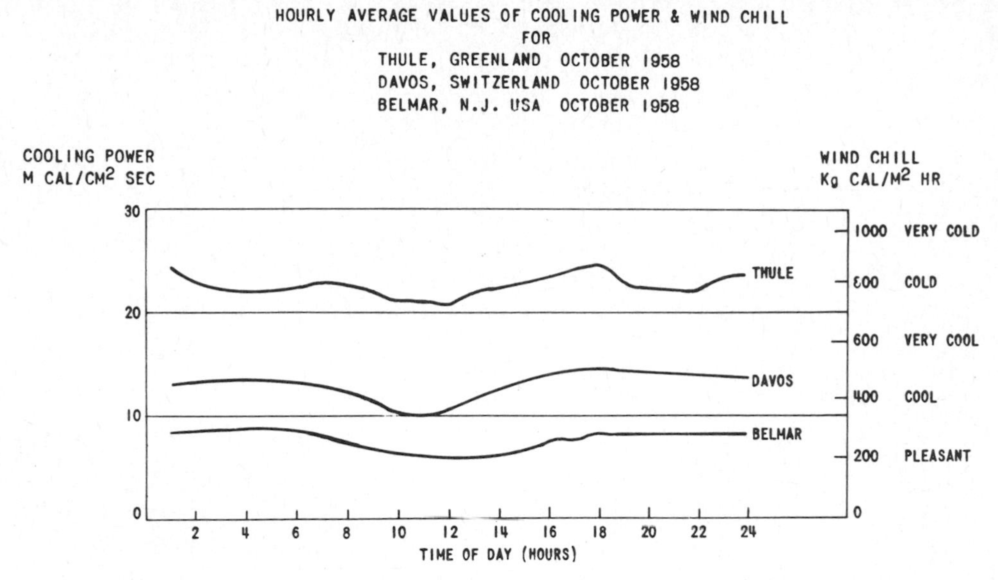

Figure 2.8. Diurnal variation of wind chill and cooling power in three different environments (from work by Flach and Mörikofer, 1962-67)

The first term represents the wind influence, which in this and similar representations shows that the square root of the wind speed governs the effect. The second term represents the difference between body and air temperature.

The wind chill equivalent temperature, which is broadcast in winter by many radio and television stations, is a valuable aid for persons exposed to a cold outdoor environment to choose appropriate clothing. This can prevent frostbite, disease or even death (Falconer, 1968). Much effort has been spent to design adequate protective clothing for a variety of climatic conditions. Space permits only a very short discussion here. For elaborate details the reader is referred to books by Fanger (1970) and Hollies and Goldman (1977).

The principal objective of clothing is insulation against the effects of atmospheric conditions on the human heat balance. It is essentially an artificial device to maintain homeostasis. Clearly the main problem is to guard against excessive heat loss. This is achieved by fabrics which incorporate air and keep stagnant air layers near the body. Air is an excellent insulator. The resistance to heat transfer is given by a dimensionless unit reflecting the insulation of clothing:

$$(7) \qquad I_{cl} = \frac{T_s - T_a}{Q_b} - \frac{I_b(Q_b - Q_e)}{Q_b}$$

where I_{cl} = insulation of clothing
T_s = skin temperature (in equilibrium assumed to be 33°C)
T_a = air temperature, °C

I_b = resistance to heat transfer of boundary layer of air at surface of clothing. ($I_{cl}+I_b$ give resistance to both radiative and convective heat transfer)

Q_b = heat flux from body (generally set at constant value of 75 per cent of metabolic rate)

Q_e = radiative heat flux from environment (outdoors principally the direct solar and sky radiation)

For practical purposes a unit has been introduced which represents the amount of insulation which will permit the flux of 1 kcal m^{-2} h^{-1} through a garment with a temperature difference of 0.18°C between the inner and outer surface of the fabric. This unit has been dubbed *clo.* It is the amount of insulation needed by a person sitting quietly in a room at 21°C with an air flow of $u < 5$ cm sec^{-1}. By these definitions

$I_{cl} = \frac{R_{cl}}{0.18}$ Clo. The value of the quantity of the boundary layer of air insulation is generally set at

$$(8) \qquad I_b = \frac{1}{0.62+0.19u^{0.5}} \text{ clo}$$

where w = is air motion in m sec^{-1}.

Another unit has also been used occasionally. It is called a tog, which is about 0.645 clo.

Substituting the value of I_{cl} in formula (7) a combined approximation results (de Freitas, 1979):

$$(9) \qquad I_{cl} = \frac{33-T_a}{0.155Q_b} - \frac{Q_b+Q_e}{(0.62+0.19u^{0.5})Q_b}$$

Air temperature and wind speed are readily obtained and the metabolic heat flux can be quite well estimated from a person's activity level. The only element which is more difficult to enter is the environmental heat flux. Usually only the sun and sky radiation on a horizontal surface is available from meteorological observations. That leaves the value of the reflected and infrared radiations usually open. In particular, there is generally very little information on the nocturnal atmospheric counter-radiation. Hence calculations of the clothing required according to formula (9), are difficult but a good guess can be made. The proper choice can be made by knowledge of the insulating values of various fabrics. From a number of sources and many chamber experiments the information presented in Table 2.7 has been gathered.

If this table looks biased towards male attire, it must be noted that the tests leading to this clothing classification were mostly made on military uniforms. For very cold weather designs of outer-wear jackets of quilted, fibre-filled material have become in recent years popular with both sexes and the last two categories of the table are not much different for male or female. Specially designed thermal underwear is readily available for cold regions. Military tests have also shown which environmental temperatures are considered comfortable with clothing of 1 clo at various metabolic rates, as shown in Table 2.8.

Table 2.7. Insulation value of various garments

Type of clothing	clo value
Tropical: open-neck shirt, shorts, sandals	0.1
Light summer wear: open-neck shirt, slacks, ankle socks, shoes	0.3-0.4
Comfortable weather wear: business suit, short cotton underwear, socks, shoes	1
Cool weather wear: business suit, light underwear, socks, shoes, light overcoat	1.5
Cold weather wear: business suit, underwear, heavy socks, shoes, hat, overcoat	1.8-2.5
Very cold weather wear: as for cold weather plus gloves and heavy overcoat, hat	2.6-3.5
Polar region wear: woollen underwear, coveralls, parka with hood, mittens, fur-lined boots	3.6-4.7

Table 2.8. Relation of metabolic rate to environmental temperature with clothing insulation of 1 clo

Activity	Metabolic rate (MET)	Air temperature °C
Sedentary desk job	1	21
Marching with pack	3	4
Running with pack and rifle	6	–20

Hot climate problems

The challenges of cold climates can be readily overcome by proper clothing, heating of dwellings, and adequate diet; the problems of hot seasons and climates are more difficult to overcome. Even though humans are a product of the tropics, there are often forbidding atmospheric conditions which they cannot cope with physiologically. The artificial defences against heat outdoors are far fewer than those against cold. There is nothing to shield us from high temperatures, or what is worse, a combination of high temperature and humidity. There is some protection against intense short-wave solar and sky radiation by reflecting clothing or by shade-producing parasols. Indoors, of course, one can use air conditioning, which can both lower the temperature and the humidity. Artificial ventilation will also help. However, the energy costs of air conditioning are high.

The physiological defences against heat and humidity have been well explored and a large literature exists (Ladell, 1957; Sulman, 1976). Again the body will try to maintain its core temperature. An internal temperature rise will trigger the heat-regulatory function of the hypothalamus gland. At first this will lead to vasodilation in the extremities, especially hands and feet. The amount of blood flowing through the

Table 2.9. Wind-chill equivalent temperature

Wind velocity (mph)	(m sec^{-1})	Dry-bulb ambient temperature (°F and °C) 50 (10.0)	41 (5.0)	32 (0.0)	23 (–5.0)	14 (–10.0)	5 (–15.0)	–4 (–20.0)	–13 (–25.0)	–22 (–30.0)	–31 (–35.0)	–40 (–40.0)	–49 (–45.0)	–58 (–50.0)
		Equivalent temperature (°F and °C) (equivalent in cooling power on exposed flesh under calm conditions)												
Calm	Calm	50 (10.0)	41 (5.0)	32 (0.0)	23 (–5.0)	14 (–10.0)	5 (–15.0)	–4 (–20.0)	–13 (–25.0)	–22 (–30.0)	–31 (–35.0)	–40 (–40.0)	–49 (–45.0)	–58 (–50.0)
5	2.2	48 (8.9)	38 (3.3)	27 (–1.7)	20 (–6.7)	10 (–12.2)	1 (–17.2)	–9 –22.8)	–18 (–27.8)	–28 (–33.3)	–37 (–38.3)	–47 (–43.9)	–56 (–48.9)	–65 (–53.9)
10	4.5	40 (4.4)	29 (–1.7)	18 (–7.8)	7 (–13.9)	–4 (–15.6)	–15 (–26.1)	–26 (–32.2)	–37 (–38.3)	–48 (–44.4)	–59 (–50.6)	–70 (–56.7)	–81 (–62.8)	–92 (–68.9)
15	6.7	36 (2.2)	24 (–4.4)	13 (–10.6)	–1 (–18.3)	–13 (–25.0)	–25 (–31.7)	–37 (–38.3)	–49 (–45.0)	–61 (–51.7)	–73 (–58.3)	–85 (–65.0)	–97 (–71.7)	–109 (–78.3)
20	8.9	32 (–0.0)	20 (–6.7)	7 (–13.9)	–6 (–12.1)	–19 (–28.3)	–32 (–35.6)	–44 (–42.2)	–57 (–49.4)	–70 (–56.7)	–83 (–63.9)	–96 (–71.1)	–109 (–78.3)	–121 (–85.0)
25	11.2	30 (–1.1)	17 (–8.4)	3 (–16.1)	–10 –23.3)	–24 (–31.1)	–37 (–38.3)	–50 (–45.6)	–64 (–53.3)	–77 (–60.6)	–90 (–67.8)	–104 (–75.5)	–117 (–82.8)	–130 (–90.0)
30	13.4	28 (–2.2)	14 (–10.0)	1 (–17.2)	–13 (–25.0)	–27 (–32.8)	–41 (–40.6)	–54 (–47.8)	–68 (–55.6)	–82 (–63.3)	–97 (–71.7)	–109 (–78.3)	–123 (–86.1)	–137 (–93.9)
35	15.6	27 (–2.8)	13 (–10.6)	–1 (–18.3)	–15 (–26.1)	–29 (–33.9)	–43 (–41.7)	–57 (–49.4)	–71 (–57.2)	–85 (–65.0)	–100 (–73.3)	–113 (–80.6)	–127 (–88.3)	–142 (–96.6)
40	17.9	26 (–3.3)	12 (–11.1)	–3 (–19.4)	–17 (–27.2)	–31 (–35.0)	–45 (–42.8)	–59 (–50.6)	–74 (–58.9)	–87 (–66.1)	–102 (–74.4)	–116 (–82.2)	–131 (–90.6)	–145 (–98.3)
45	20.1	25 (–3.9)	11 (–11.7)	–3 (–19.4)	–18 (–27.8)	–32 (–35.6)	–46 (–43.3)	–61 (–51.7)	–75 (–59.4)	–89 (–67.2)	–104 (–75.5)	–118 (–83.3)	–133 (–91.7)	–147 (–99.4)
50	22.4	25 (–3.9)	10 (–12.2)	–4 (–20.0)	–18 (–27.8)	–33 (–36.1)	–47 (–43.9)	–62 (–52.2)	–76 (–60.0)	–91 (–68.3)	–105 (–76.1)	–120 (–84.4)	–134 (–92.2)	–148 (–100.0)
		Little danger				Increasing danger			Great danger					
		Danger from freezing of exposed flesh (for properly clothed persons)												

Note 1. – To temperature reproduced originally in °F, corresponding values in °C in brackets are added.
Note 2. – For wind values of $\leqslant 1$ m sec^{-1}, conditions are assumed to be calm.
The table indicates the limits of danger of frostbite even for appropriately dressed persons.

peripheral skin vessels can increase by an astounding factor of 7, compared to non-stressful thermal environment. This causes a rather dramatic increase in heartbeat rate if the core (rectal) temperature should increase by only 1°C (from normal 37° to 38°C). For a resting person the pulse rate may increase by 30 per cent, for a working person it can go up by 40 per cent. Very quickly hypothalamic regulation will initiate sweating. This too can reach extraordinary values, maximally as much as 1.7 litres per hour, or up to 10 litres in an 8-hour work day in the tropics. The increase in peripheral blood flow and skin temperature will cause greater heat loss by radiation from the body and by convection. Sweating will cause evaporative heat loss which will be assessed more quantitatively below.

The consequences, if the core temperature rises are very dangerous. They can lead to heat syncope (unconsciousness) because the brain does not get enough blood as a result of the peripheral vasodilation. Should the core temperature rise to about 42°C heat stroke (hyperpyrexia) will occur, characterised by a complete collapse of the cardiovascular system, often with fatal results. Excessive sweat production can result in dehydration and salt depletion. The loss of body fluids can lead to giddiness and ultimately delirium. The salt depletion will cause heat cramps, nausea, and vomiting. Unless adequate fluids are consumed and the electrolyte balance is restored death may result (Gilat *et al.,* 1963).

Atmospheric heat conditions are particularly dangerous to infants, whose heat regulatory mechanisms are not yet developed, and to old persons whose circulatory systems are impaired. Obese individuals also are endangered by heat. Persons with congestive heart disease are also among potential victims of hot weather or climates.

There are again a large number of indices attempting to assess the heat stress of the environment. These relate to the feeling of discomfort and, as such, have been frequently classified by votes of people exposed to a variety of environmental conditions. One of these indices which has been widely used is the so-called effective temperature. It is in fact a combination of air temperature and humidity into an equivalent temperature representing the prevailing conditions, the same sensation as felt for a temperature of the same value with calm air saturated with water vapour. It can be represented by

$$ET = 0.4\ (T_d + T_w) + 4.8 \qquad (10)$$

where T_d = air temperature measured with °C dry-bulb thermometer
T_w = temperature measured with a moistened (wet) thermometer. This is $\leqslant T_d$ and the difference $T_d - T_w$ is a measure of atmospheric moisture. The greater the difference the drier the atmosphere, because of the evaporation from the wet wick surrounding the thermometer, the heat of vaporisation being taken from the thermometer.

Figure 2.9 gives a comfort diagram based on temperature and humidity factors. It shows in a grid of dry-bulb air temperatures and water vapour mixing ratios (grams of water vapour per kilogram of dry air), lines of wet-bulb temperatures, relative humidities, and effective temperatures. A hatched area in the graph indicates the combination of values within which a majority of persons in light clothing and sedentary occupations feel comfortable. This clusters around effective temperatures of 20°C (or dry-bulb temperatures 20 to 24°C and relative humidities of 40 to 60 per cent). At effective temperatures above 30°, far into the zone of discomfort, the dangerous area of core overheating begins, and at 32°C effective temperature the physiological defences begin to break down, as shown in Figures 2.10 and 2.11. Hard work is not recommended for any effective temperature above 25 to 28°C, depending on degree of acclimatisation. The wise custom of a siesta during the hottest hours of the day in tropical and subtropical climates is clearly supported by climatic stress analysis.

The effective temperature concept was developed for indoor conditions and for assessment of needs for air-conditioning. For outdoor conditions one of the major modifications is the solar radiation load which in experiments (Lee and Vaughan, 1964) of persons exposed to full summer sun is about equivalent to a 7 to 9°C increase in air temperature.

Because of the great importance of heat stress in exposed occupations, a special piece of equipment is used which measures both the dry-bulb and wet-bulb temperature, and at the same time integrates the effect of radiation and wind on a black globe, measuring its temperature, Tg, too. A combination of these temperatures

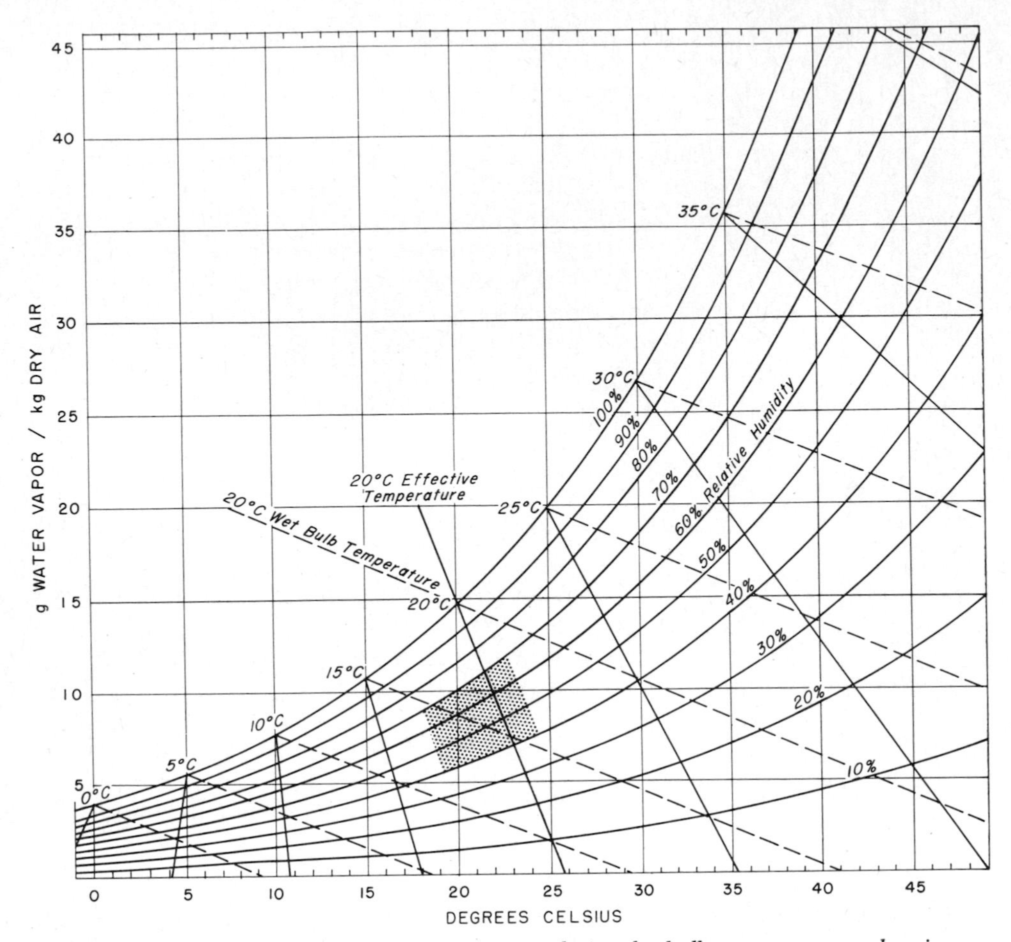

Figure 2.9. Thermodynamic comfort diagram, relating dry bulb temperature and various humidity measures (vapour pressure, wet bulb temperature, relative humidity) to effective temperature. Shaded area is the zone of comfort for the majority of persons.

is called the wet-bulb globe temperature *(WBGT)*, with the following partitioning:

$$WBGT = 0.2T_g + 0.1T_d + 0.7T_w \quad (11)$$

This indicates the great weight one has to attribute to the moisture-indicating wet-bulb temperature. For *WBGT* values of 31°C or higher it has been found wise to discontinue strenuous activities.

For the lay public indices such as effective temperature or wet-bulb globe temperatures have much less appeal than for scientists or engineers. A particular drawback is the fact that the uncomfortable or dangerous values are lower than the air temperature. Hence the Canadian Weather Service developed an index, in analogy to the equivalent wind chill temperature, which would be a better comfort (or

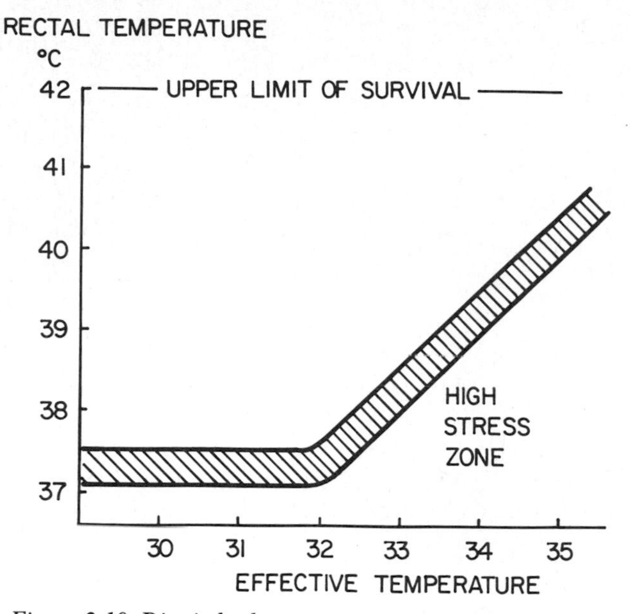

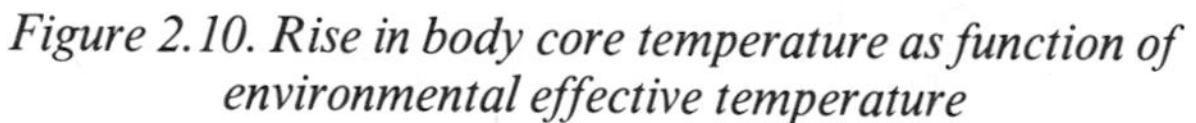
Figure 2.10. Rise in body core temperature as function of environmental effective temperature

Figure 2.11. Combinations of air temperature and relative humidity conducive to heat stroke

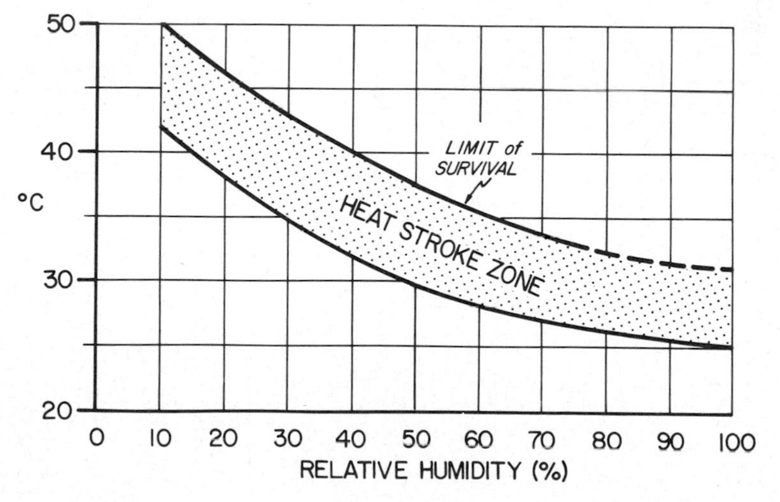

discomfort) representation of summer weather (Masterson and Richardson, 1979). This index, called humidex, is defined as

$$(12) \qquad H = T_a + h$$

where T_a = air temperature
h = 5/9 (e–10)
e is the vapour pressure in millibars (or kilopascals × 10).
e is determined from T_{dp} or dew point temperature, i.e. the temperature to which the air has to be cooled for dew to form.
T_{dp} is another measure of atmospheric moisture.

$$(12a) \qquad e = 6.11\left[\exp\left\{\frac{ML}{R}\left(\frac{1}{273.16}-\frac{1}{T_{dp}}\right)\right\}\right]$$

where M = molecular weight of water (18.016 g mol^{-1})
L = latent heat of vaporization (597.3 cal g^{-1})
R = gas constant (8.3144 × 10^7 erg mol^{-1} sK^{-1})
T_{dp} = dew point temperature, °K.

It is fairly clear that appropriate tables will greatly facilitate by-passing the fairly complex calculation to obtain H. Table 2.10 gives a brief excerpt from such tables.

Where dew point temperatures, T_{dp}, are available the value of h can be added to the air temperature. The conversion value has been determined by Canadian researchers, as shown in Table 2.11.

The degree of discomfort for various ranges of the Humidex have been described. They are shown here in Table 2.12.

While such indices may satisfy certain practical needs, for scientific work in human physiology and for quantitative estimates of energy consumption for air-conditioning

Table 2.10. Example of Humidex values as function of air temperature and dew point

Air temperature	Wet-bulb temperature T_w, °C		
T_{dp}, °C	15	20	25
20	22.3	27.6	—
25	25.5	30.8	37.2
30	—	34.0	40.4
35	—	37.2	43.6

Table 2.11. Conversion value of dew point temperature to factor h in Humidex formula

T_{dp} °C	15	20	23	25	26
h	4	8	10	12	13

Table 2.12. Comfort sensations associated with various ranges of Humidex

Humidex value	Sensations
20-29	Comfortable
30-39	Various degrees of discomfort
40-45	Almost everyone uncomfortable
$\geqslant 46$	Work effort must be restricted

a more exact approach is needed. This can be done by an energy balance approach (Auliciems & Kalma, 1979), which defines the heat stress, *HS,* as follows:

$$(13) \qquad HS = (M+(1/A)(Q_s+Q_{sk})(A/f)$$

where M = metabolic rate
A = body area
Q_s+Q_{sk} = solar and diffuse sky radiation
f = cooling efficiency of sweating,
l/c = $\exp(0.6\frac{E}{E_{max}} - 0.12)$
C = actual sweating rate, $E = AM + (Q_s+Q_{sk}) - AD$:
D_i = dry heat exchange loss at 35°, when sweating = $h\,(35-T_a)$;
h = heat transfer coefficient from body to air.
E_{max} = evaporative capacity of the air = $49.61u^{0.3}(31.5\ e)$ watts;
[u = wind speed m sec^{-1}, e vapour pressure (millibars)]

The investigators, by substituting the various heat gain and heat loss estimators, worked out a heat stress relation:

$$(14) \qquad HS = [132+(Q_sQ_{sk})-(12.3+16.2u^{0.5})(35-T_a)\ \exp 0.6(\frac{E}{E_{max}} - 0.12)]\text{Watts}$$

using all previously explained symbols. In order to perform the necessary calculation, air temperature, wind speed, vapour pressure, solar and sky radiation must be known. Metabolic rates and body area can be assumed for various individuals and work loads. Using available climatic data, Kalma and Auliciems (1978) charted heat stress in Australia for various seasons and day times. They also estimated both heating and cooling energy requirements.

The need for air-conditioning in hot climates has been investigated in various contexts. Among the more interesting are performance and learning. Both suffer in hot environments. Tasks requiring memory are impaired also when the thermal equilibrium is disturbed. The evidence is overwhelming that the mental balance is upset by heat discomfort also. When persons cannot get rid of their metabolic heat they became irritable. Tempers flare, crimes of passion rise in frequency in hot weather and riots are more common than in comfortable weather. Hot desert winds, often laden with dust, whether called Chamsin, Sharav or Scirocco, seem to increase human aggressiveness.

But heat waves are not only reflected in the crime statistics; they also show clearly in the mortality statistics. About 500 persons per year die on average in the United States directly as a result of heat stroke. In 1936, one of the hottest summers on record, especially in the Midwest and Great Plains, 4,700 people were killed by the heat. These are only the direct victims. Many more die each year from other causes aggravated by excessive heat. The year 1966 was another of high mortality in the US coincident with high temperatures. St Louis and New York City showed, for example, suddenly high mortality rates. Figure 2.12 shows the weekly death rates for New York City during the summer of 1966. The hatched band shows the 95 per cent confidence limits of expected deaths, based on long-year experience. The dashed line represents the weekly mean temperatures, which rose in the last June week to 30°C. This was followed in the subsequent week by a 35 per cent rise in deaths, far beyond the expected values. Ailing elderly persons were the chief victims.

It is interesting to note that the upper confidence limit shows a notable bulge in June. This suggests that during that month somewhat higher death rates are not uncommon. In this transition from spring to summer people are not yet properly acclimatised to heat, and their clothing had not been adjusted to warm weather.

In 1976, London, England, experienced a similar startling rise in summer deaths among the elderly as had New York a decade earlier. The hot spell from June 23 to July 7 was the worst in the century. Temperatures were 10°C above the long-term average and maxima soared above the 30°C mark. Death rates among the population of 65 years or older rose steeply by 50 per cent.

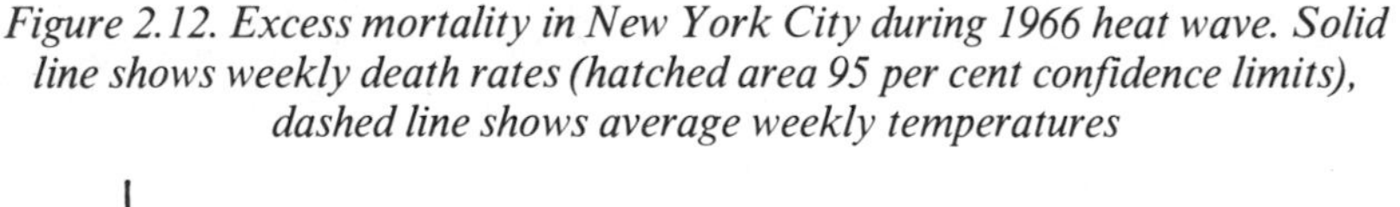

Figure 2.12. Excess mortality in New York City during 1966 heat wave. Solid line shows weekly death rates (hatched area 95 per cent confidence limits), dashed line shows average weekly temperatures

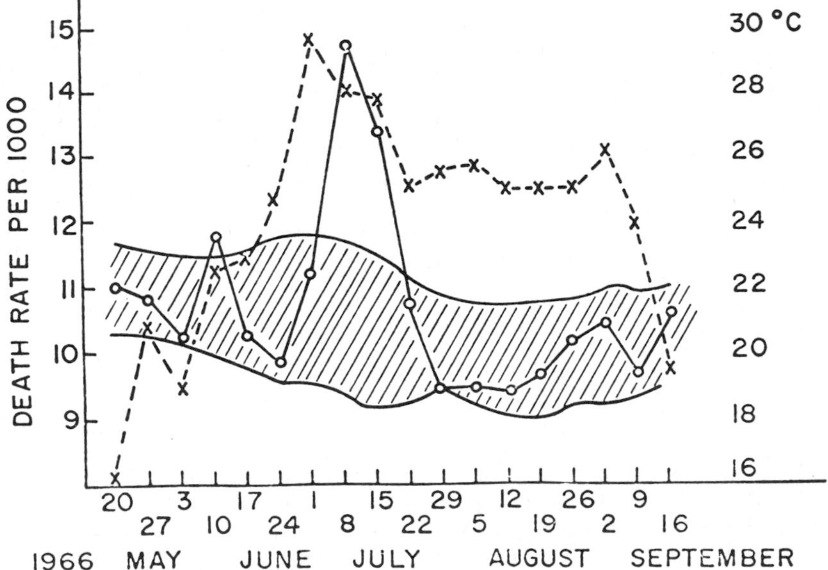

There have been a number of attempts to relate incidence of criminal offences to weather and climate. Whatever the relationship may be, they are neither obvious nor simple. If the statistics show that there are more robberies in the cold season, does that indicate anything more than that the longer period of darkness offers better cover than short nights? On the other hand, aggravated assault and rape are more frequent in the warm season. The traditional folklore has it that hot weather leads to so-called crimes of passion. But some criminologists have raised the question whether the victims are in the season less careful and thus more exposed to criminal attacks (Kevan and Faust, 1976).

Accident statistics also suggest that hot weather renders many more accident prone. It has been argued that excess heat makes people sleepy and hence less alert. Thus both industrial and traffic accidents show increases in hot weather for no other obvious reasons, that are quite apparent when roads are slippery from rain or ice.

CONTAMINANTS OF THE ATMOSPHERE

The air, far from being a pure mixture of the life-sustaining gas oxygen (O_2) and the inert gas nitrogen (N_2), contains innumerable gaseous and solid admixtures. Some of these are harmless, others noxious. Many of them are produced by nature, but human activity has progressively added more and more. Amounts of these contaminants vary greatly with geography and climate. In urban and industrial areas they have become major elements of the local climate.

Among the natural contaminants are dusts from deserts and volcanic eruptions. The former are a regular feature in arid countries. Blown by the wind they can travel thousands of kilometres. For example, Sahara dust is a regular feature in the Caribbean and major dust clouds can be frequently noted over the South Atlantic on satellite pictures. Volcanic dust is more sporadic and generally restricted to the areas of active volcanoes. But the volcanoes contribute also to the sulphur dioxide (SO_2) content of the air. Other sulphur compounds, such as the evil-smelling hydrogen sulphide (H_2S) result from organic decay. But manmade SO_2 has been blamed for mortality and morbidity in major air pollution episodes. In many of these, when SO_2 exceeded 0.2 parts per million, acute respiratory distress ensues. It may result in acute bronchitis and pneumonia. Persons with existing respiratory ailments, such as emphysema, may die during a persisting air pollution episode.

These episodes are caused by meteorological conditions. They always involve a temperature inversion, in which a shallow layer of cold air is overlaid by warm air. This condition occurs in high pressure systems with often clear skies. Winds are weak and pollutants brought into the air from whatever source will tend to accumulate. Inversions are frequent in the valleys of mountainous areas and common in some subtropical coastal zones. Often geography and meteorology conspire to create stagnant conditions. Mountain barriers may block ready air exchange. The Los Angeles basin is a typical example.

In urban areas where pollution is endemic, chronic bronchitis, emphysema, and asthma are common ailments. In the most affected people continuous exposure may result in pulmonary fibrosis or emphysema. It is, however, difficult to relate death rates to the level of pollution, except for the major pollution episodes, because of the many other insults, such as cigarette consumption, and the great variability of pollutants even in different city quarters (Lipfert, 1980). Vital statistics tell us very little about shortening of life.

Air chemistry is exceedingly complex. No contaminant stays unchanged. With time and such energy input as solar radiation, photochemical reactions take place and create new compounds. Trace metals often act as catalysts in the promotion of chemical reactions. Thus SO_2 and NO_2, common in the smoke plumes of power plants, other industrial processes using fossil fuels, and car exhaust, are converted to sulphuric acid (H_2SO_4) and nitric acid (H_2NO_3). Photochemical reactions will produce from NO_2, nitrous oxide (NO) and O, which in turn produce ozone O_3, and peroxyacetylnitrate (PAN). The latter is a well-known eye irritant. Ozone in concentrations of 0.1 parts per million will lead to irritation of mucous membrances and some persons experience nausea and headache.

Another contaminant, prevalent in cities but encountered only in extremely low concentrations in nature, is carbon monoxide (CO). The culturally produced CO results from incomplete combustion of fossil fuel, most of it from car exhaust, but is also present in smoking material. Because of its affinity for the haemoglobin in blood it forms a compound carboxyhaemoglobin (COHb) and the blood molecules so transformed become useless for the essential transport of oxygen to muscle and vital organs. The blood transformation is directly proportional to the CO concentration. In 10 parts per million (rare in the atmosphere) 1.6 per cent of the haemoglobin is transformed into COHb.

Among the solid suspensions, often designated as aerosols, nature produces a wide share of particles that cause distress. Many of these are aeroallergens, with the most common being various types of pollen. For nearly the whole vegetative period of the year some of this flotsam is present in the air. In the spring tree pollen is prevalent and grass pollen is usually also common in that season in moderate latitudes. In many areas goldenrod and rag weed pollen saturate the autumn air where these plants grow. At the time of maximum production of a given pollen variety one can count from 10 to 300 such particles in a cubic metre of air. They are relatively large, about 30 μm in diameter and are intercepted in the outer respiratory passages. Yet they cause a large amount of misery, from runny noses and sneezing to asthma.

Pollen are larger and less numerous than other aerosol particles. These measure usually from 0.1 to 10 μm. Larger particles fall out close to the source, but the small ones are carried by the wind. Rainfall causes their release from the plants. Depending on the climatic rain regime their life time as atmospheric suspensoids may be from 5 to 30 days. The smallest variety is also the most numerous. Least are found over the ocean, where there are only a few hundred per millilitre of air (ml). In clean country air there are usually several thousand, and in city air less than 100,000 ml^{-1} are rarely encountered. While ocean air is unlikely to contain noxious substances and may, in fact, carry some desirable ones such as iodine, urban air has a large number of undesirable trace substances. Among them are asbestos, a carcinogenic mineral, cadmium, detrimental to the cardiovascular system, and others.

It should be noted here that the average human being circulates about 12 m^3 air through the lungs daily. This means that a staggering 10^{12} particles per day pass through the respiratory system of the average urban dweller. Many are exhaled, some are intercepted in the upper respiratory passages but many penetrate into the alveoli. It is known that on autopsy lungs of urbanites are dark, instead of pink, as found in the lungs of people living in clean air.

At this point we should look at the portion of the atmospheric contaminants which are radioactive. Here again nature exposes us to considerable radiation. Two gases Radon (Rn^{220}) and Thoron (Th^{222}). Both of them are decay products of the naturally occurring radioactive elements in the early's crust. They escape through the soil into the air where they continue to decay. In areas of granitic rock much more radioactive gas is emitted than from thick sedimentary layers. Radon has three radioactive solids as decay products and thoron six. These are part of the aerosols. They are almost absent over the oceans and a snow cover will prevent the diffusion of radon and thoron from the soil.

The radiation is about 0.16 nano-Curies[4] from each of the two radioactive gases. It is not known whether there is any repair in cells subjected to very minor doses of radiation or if the detrimental effects are cumulative. The biological effectiveness of these natural exposures are measured in a unit called a RAD (radiation absorbed dose), which is the energy absorption of an energy of 100 erg per grams of substance. It is estimated that most people are exposed to from 50 to 150 m rad yr^{-1}. But there is also an exposure to manmade radioactivity. Apart from weapon testing there is a small addition to the air from the nuclear and coal-fired power plants. The latter, using certain types of coal which contain uranium and thorium often release more radioactivity than nuclear fission plants, which are more closely controlled. Present (1980) exposure to manmade radioactivity of all types, including weapons, has been estimated at about 3 per cent of the natural radiation components.

Atmospheric radioactivity and cosmic radiation causes ionisation of air molecules and these air ions have been adduced as the cause of innumerable health effects. Observations show that there are usually between 500 and 2000 ions cm^{-3} in air. The positive charge prevails over the negative charge by a ratio of 5 to 4 (Wehner, 1969). There is a continuous recombination into the electrically neutral state. Many of the ions attach themselves to other aerosol particles. They then become inert and recombination with particles of opposite charge is slowed down. The initial ions are called small ions. They have high mobility in an electric field. Those attached to aerosols are called large ions and have slow mobility in an electric field.

There is some evidence that in certain weather conditions their concentration rises above the average. An example is the conditions preceding or accompanying the hot desert wind Sharav. Sulman *et al.* (1974) reported in Jerusalem a quadrupling of the ion number (4000 – and 4500+) prior to such winds. The adverse reactions of weather sensitive persons were attributed to "serotonin irritation syndrome." The neurohormone serotonin is an irritant. It inactivates an enzyme (monoaminooxydase) and reduces its metabolite 5-hydrooxindole. Excessive serotonin causes rise in blood

4. 1 nano-Currie corresponds to 37 radioactive disintegrations per second. Both Rn and Th decay by α-emissions.

pressure, as well as irritability and allergy symptoms. It is alleged that the offending serotonin is secreted in the intestinal tract and from there is pervasively transported through the body.

In the case of the Föhn,[5] a hot wind descending into the valleys when air masses are forced to cross a mountain range and in the process lose their moisture on the windward side, it is well documented that the surface radon content increases. It often marks the transition from anticyclonic to cyclonic circulation in the Alpine areas. Physiological symptoms in weather-sensitive persons are inability to concentrate, motor-unrest, sleep disturbances, occasionally migraine headaches, but occasionally result in euphoria (Swantes and Reinke, 1978). If all this can be ascribed to ionisation, increase or the prevalence of the positive sign over the negative is not yet adequately established, although the adherents of the serotonin hypothesis suggest that.

DISEASES AFFECTED BY CLIMATE

An examination of maps showing the geographical distribution of diseases clearly shows that pathological factors are quite unevenly distributed over the globe. The prevalence of certain diseases is so notable that they have been labelled "tropical". Although there are cultural, nutritional, and sanitary facets of disease occurrence and control, it is immediately clear that climate and weather play an important role (May, 1950; Stamp, 1964). Although the tropics have been singled out as the hotbed of a host of diseases, there are other diseases which are more prevalent in moderate or cold climates.

The association of various diseases with climatic factors is generally established only in a statistical sense. In many instances the exact pathway of the etiology is not well known. What complicates the tracing of causes and effects is the fact that climatic circumstances will not only affect the human body but also the microorganism or viruses causing the disease. However, in very many instances the effect is on intermediary hosts and transmitters. These vectors are almost always weather-sensitive at various stages in their life cycle. This accounts also for the very notable annual variation of diseases, especially at onset.

In those diseases where insects or snails act as vectors the limiting climatic conditions are fairly easily traced by comparisons of maps of disease distribution and climatological charts (Jusatz, 1962; Jusatz, 1977). This is particularly evident for the various disease-transmitting mosquitoes. Of all climate-affected vectors, the mosquito is by far the most noxious (Gillett, 1974). Of 3,000 mosquito species, about 40 to 50 are important as malaria transmitters. Their life cycles are intimately related to weather. Optimal for oviposition are, for example, temperatures of 30°C. Annual temperatures below an average of 12°C are generally limiting for survival. Ample rainfall for the production of stagnant surface waters is essential to keep the larvae alive. There is also evidence that the malaria plasmodium requires temperatures of the warmest month of

5. In the North American Rocky Mountains this type of wind is known by its Indian name Chinook.

the year to be greater or equal to 17°C. There is little doubt that malaria is the most prevalent disease afflicting humanity. There are estimates that 100 million cases occur annually and result in a million deaths (National Academy of Sciences, 1973).

Yet there are a large number of other diseases transmitted by insects such as Leishmaniasis (sandfly), sleeping sickness (tsetse fly) and yaws, which can also be transmitted from person to person. For nearly all disease transmitting bloodsucking insects the 10°C annual isotherm is the lower limit of viability.

Other insects important as vectors of diseases are ticks and fleas. Ticks, which are important in the transmission of meningitis, are generally hardy, although for hatching they need temperatures greater or equal to 22°C. Fleas ordinarily survive on warm-blooded secondary hosts, such as dogs, cats or rats, so that their sensitivity to environmental conditions is sharply reduced. Table 2.13 shows some of the established limiting climatic conditions for certain diseases or their vectors.

Other tropical diseases are transmitted by snails, such as schistosomiasis which is fairly widespread. The intermediate hosts, which carry the blood flukes causing the disease, are found in an aquatic environment. Hence ample rainfall feeding streams, swamps or puddles are essential climatic prerequisites.

There are some diseases which evidently do not follow the moist-warm environmental pattern otherwise typical of tropical diseases. Jusatz (1962) reported that the often epidemic cerebro-meningitis in the sub-Saharan area is bound to the dry season, January to April, when the notorious dry, dusty Hamattan wind from the desert sweeps the area. The epidemics cease with the start of the rainy season caused by the northward motion of the intertropical convergence zone.

In the eastern United States, however, epidemic encephalitis is invariably tied to wet weather. It can spread as far as southern New England. Hence the larvae of the transmitting mosquito *(Culiseta melanura)* must be fairly resistant to cold. Outbreaks of the disease are bound to years in which the preceding August to October had heavy precipitation, often caused by tropical storms. During the epidemic year June through August rains also have to be unusually heavy (Hayes and Hess, 1964). In Europe, acute cases of meningitis epidemica have been observed to occur with higher frequency when weather changes indicating the approach of a low pressure system (increasing cloudiness, increasing temperature and humidity, falling barometer) occur (Brezowsky and Menger, 1959).

Table 2.13. Climate dependence of selected diseases

Disease	Limit °C	Vector	Limits °C	
Leishmaniasis	Annual ⩾ 10	Sandfly	Annual ⩾ 10, low humidity	
Dengu fever	Annual ⩾ 12	*Aedes aegypti*	Annual ⩾ 12	
Sleeping sickness	Warmest month ⩾ 25	Tsetse fly	Min ⩾ 10	
Yellow fever	Coldest month ⩾ 15	*Aedes aegypti*	Annual ⩾ 12	large amounts of rain
Malaria	Annual ⩾ 15	Anophelidae	Annual ⩾ 12	
Amoebic dysentery	Warmest month ⩾ 25	*Entamoeba histolytica*	Insensitive	

Another tropical disease, classified among the rheumatic ailments is myositis tropica, an inflammation of the muscle fibres. It is common in the tropical rain forests and occurs in the savannah during the rainy season but is rare in the warm deserts.

In contrast, scarlet fever is a typical disease of the cold-moist regions of the world (von Bormann, 1960). In continental climates incidence of the disease rises steeply in the cold months and drops precipitously in the warm months. In oceanic cool climates the annual variation is quite smooth and the maximum may be reached at the beginning of the cold season in October. Epidemic outbreaks have been noted in cold winters. In the tropical rain regions and deserts this disease is almost unknown.

In some diseases a meteorological connection can only be suspected because of the annual variation. Thus mortality from certain diseases shows distinct seasonal peaks (Brezowsky, 1964; Haase and Leidreiter, 1975). This is particularly pronounced in cases of cardio-vascular and respiratory diseases. In many areas of the world with moderate or cool climates these show notable cold season peaks and one can see an inverse correlation to the annual march of temperature (Figure 2.13). The amplitude of respiratory disease mortality is much greater than that for cardiovascular disease. This has been ascribed to the greater exposure to contact infection in enclosed areas during the heating season. Meteorologically it must be considered also that physiological stresses are induced by the great temperature contrast indoors-outdoors to which many persons are exposed, the very low humidities indoors which dry the mucous membranes of the respiratory system, and the frequent abrupt weather changes during winter. This variability is clearly shown for the average day-to-day changes of the maximum daily temperatures during the month of February over the contiguous United States (Figure 2.14). Clearly certain areas, especially in the Central Great Plains and Mid-Atlantic States, are subjected to many rapid air mass changes

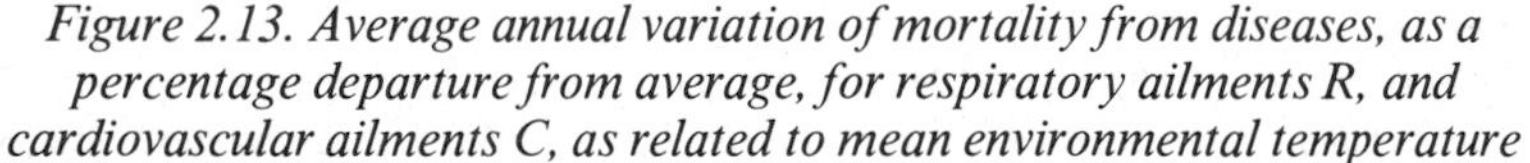

Figure 2.13. Average annual variation of mortality from diseases, as a percentage departure from average, for respiratory ailments R, and cardiovascular ailments C, as related to mean environmental temperature

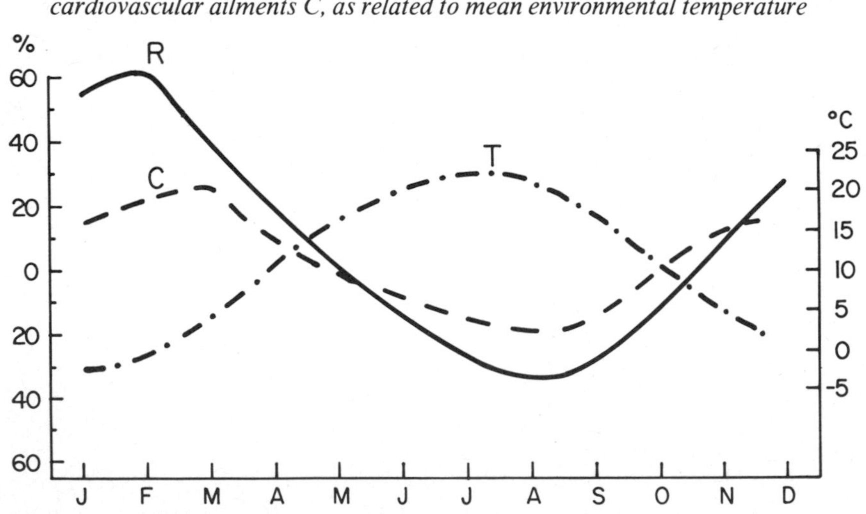

Figure 2.14. Mean interdiurnal change of daily maximum temperature in degrees Centigrade in February over the contiguous United States

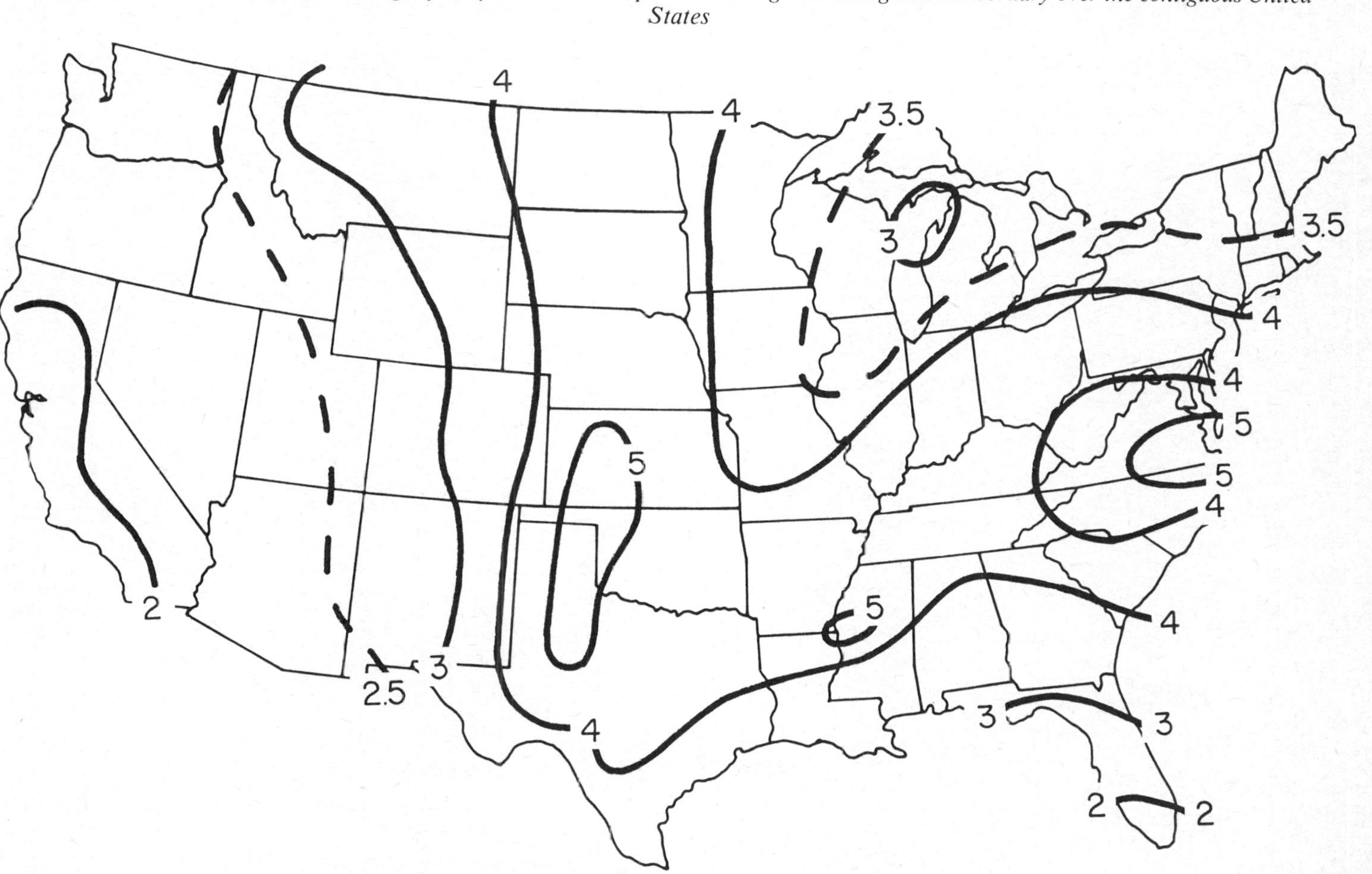

with frequent frontal passages. In regions such as these persons with chronic ailments are subjected to atmospheric changes faster than they can adapt. This applies particularly to the older age groups. A number of studies in Central Europe suggest that large-scale circulation systems — passage of marked cyclones and troughs — are associated with increased numbers of coronary occlusions and strokes (Jäger, 1968).

A disease symptom traditionally associated with weather and climate is arthritic pain. Persons especially afflicted with arthritic joint disease appear to react adversely to cold. The incidence of rheumatism and arthritis is about four times as great in a cold-moist climate as in a warm-dry subtropical area. The pain-causing sequence of events has not yet been unravelled but both statistical evidence and some chamber experiments indicate that falling pressure and increased humidity bring about swelling of afflicted joints and pain (Lawrence, 1977). One gets the impression that these joints appear to react like an aneroid barometer.

Weather as a pain-provoker is also encountered by persons with extensive scar tissue, such as those having had a limb amputated. They experience pain when atmospheric humidity rises and sometimes when temperatures change rapidly. Differential reactions of the normal skin and the scar tissue apparently lead to the pain sensations (Swantes, 1971).

One more disease should be mentioned as being weather related in some instances. That is the lung disease, histoplasnosis, which is provoked by a soil fungus. Strong winds can dislodge this organism and carry it along, exposing people to its invasion of the respiratory tract.

CLIMATIC CHANGE BY TRAVEL AND FOR THERAPY

In the modern world air travel enables people to span wide geographic areas in a very short time. The term jet-lag has become a familiar phrase. Ordinarily this refers to the upsetting of physiological circadian rhythms, such as diurnal body temperature variations, hormonal secretions, sleep patterns, and digestive processes. But often overlooked is the sudden change in climatic conditions which can be equally traumatic. Trips to the polar or tropical regions can expose a traveller to sudden temperature changes of 20 to 30°C. This is more than the change from winter to summer in many regions. The latter is a gradual transition and permits physiological adaptation. The change during a trip is abrupt in contrast. The seasoned traveller will be prepared to compensate in part for the change by bringing appropriate clothing along. Yet only the healthy will have the physiological reserve to counterbalance the climatic shock. Fortunately, there are many sources of climatological information available which can be consulted prior to embarking on a trip.

Problems are more likely to be encountered when persons from cold or moderate climates make rapid moves to warm and humid areas. The adjustments to heat stress affect the body generally more than cold because it is easier to counteract cold by clothing. Older persons in particular are adversely affected by fast change to sultry conditions. A quick trip for them may aggravate circulatory ailments and even cause heart attacks or strokes. Individuals who have slight cases of hyperthyroidism are also jeopardized (Tal and Sulman, 1973). Even healthy recreational travellers from countries with moderate climate have been warned to avoid certain areas during periods of seasonal high sultriness. Harlfinger (1975) using the readily available climatic data has pointed out the extraordinary contrast in comfort conditions that exists between the central European countries and their favorite vacation spots around the Mediterranean, especially in summer. Incipient or hidden ailments can flare acutely during sudden climatic transitions, which can expose persons also to unaccustomed foods and unfamiliar microbes.

Slower moves or moves at seasons when the climatic contrasts between two regions are minimised are usually better tolerated. There is a fair capacity for acclimatisation in humans, especially younger ones. Aside from the change in life habits, especially food and fluid secretions and of electrolyte balance, in hot regions thyroxine (a secretion of the thyroid gland) turnover is reduced, hence metabolism is slowed. This is aided by lower calorific intake. Aldosterone, an adrenal gland hormone, rises to act as a regulator of the sodium-potassium ratio, activated due to sodium loss in perspiration. Adequate intake of lightly salted fluid will help (Macfarlane, 1974). Cold acclimatisation brings about increased metabolism, usually quite voluntarily accompanied by greater calorific intake. Apparently there is also a change in the ability to shiver, a reaction to increased metabolic heat production by muscular activity. However, the ability of the human body to adapt by physiological responses to cold climatic conditions is far more restricted than to cope with warm climate. The genetic heritage of human origin in the tropics is not readily overcome (Edholm, 1966).

The role of climate in the treatment of disease has been known for a very long time. In fact one of the most common devices is to create an artificial climate to isolate a sick person from the demands of atmospheric changes. As a fairly stress-free climate, therapy has used the bed. It not only provides rest to a stricken body but permits by suitable covers or thermostatically controlled devices to regulate temperature and humidity (Spangenberg, 1954).

However, in general, when consideration is given to climate therapy it is usually an attempt to correct imbalances and gradually return bodily malfunctions back to normal. In order to enable physicians in the field of physical medicine to assess the appropriate climate for a patient a careful analysis of various climatic environments needs to be made. There have already been some promising starts in that direction. Using the available climatological data, whole areas have been charted. A notable example is Becker's (1972) system of assessment. This author distinguishes three atmospheric influence complexes: thermal, radiative, and air-chemical. Each of these is then appraised separately in three categories: stress factors, taxing factors, and protective factors.

Stress factors are:

1. high values of chill factor or cooling power; also large diurnal, seasonal, or annual fluctuations of these;
2. high intensity of radiation on a horizontal surface, including high values of ultraviolet radiation;
3. decrease of partial pressure of oxygen (elevations > 1 km);
4. intense diurnal variation of the temperature-humidity environment.

Taxing factors are:

1. temperature and humidity combinations causing sultry conditions and feelings of discomfort;
2. continued lack of radiation (including ultraviolet radiation);
3. moist-cold and frequent fog;
4. frequent or sustained high levels of air pollution either through photochemical smog or effluent accumulations under inversions and stagnant circulation conditions.

Protective factors are:

1. low values of cooling power with temperatures of 14 to 25°C, low wind speeds (< 4 m sec^{-1}), small diurnal, seasonal and annual fluctuations;
2. adequate solar and sky radiation but not too intense; shadow effects in parks and forests;
3. clean air, free of dust and pollutants from traffic or industry.

Actual meteorological values in these categories are to some extent arbitrary but some guide lines can be gathered from figures 2.5 to 2.9 in earlier sections of this chapter. Becker and Wagner (1972) have used such criteria to chart the bioclimate of the Federal Republic of Germany. The Federal Health Office of Switzerland (1965) has published, based on similar criteria, a guide to climatic resorts with indications of what ailments are thought to be beneficially influenced at various localities.

The therapeutic exploitation of various climates has been extensively explored since the last century. At one time desert climates or high altitude climates with their intense radiation and low humidities were the only remedies for pulmonary tuberculosis until replaced by chemotherapy. But even now such climatic conditions are used for recovery from respiratory ailments.

Emphysema, bronchitis and asthma can be alleviated by choice of proper climate. Certain forms of arthritis can also be beneficially influenced by a move of the sufferer to another climate. However, this is not a universal experience and individual trial periods must precede a permanent move. Circulatory ailments including high blood pressure, can be occasionally successfully treated by low-stress climatotherapy. In many instances reconvalescence from operations and severe disease is aided by a suitable sojourn in an appropriate climatic environment. Rules for climatotherapy, as a subdiscipline of physical medicine, have been worked out by physicians (Amelung, 1970; Schmidt-Kessen, 1977).

BIBLIOGRAPHY

It is clearly impossible to give within the frame of a short chapter a complete review of the vast field of bioclimatology. The interested reader is therefore referred to the more comprehensive reviews and books given below.

Becker, F. (1974) *Medizinmeteorologie, ein Grenzgebiet zur Erforschung des Einflusses von Wetter und Klima auf den Menschen.* Verein Deutscher Ingenieure, VDI-Z116, 1367-1454.

Flach, E. (1979) "Human Bioclimatology," in *World Survey of Climatology*. Edited by H. E. Landsberg, Vol. 3, Elsevier Scientific Publishing Co., Amsterdam, 1-187.

Landsberg, H. E. (1969) *Weather and Health*. Doubleday & Co., Inc. Garden City. N.Y., 148 pp.

Landsberg, H. E. (1972) *The Assessment of Human Bioclimate*. World Meteorological Organization, Technical Note No. 123, Geneva, 36 pp.

Licht, S., (Editor) (1964) *Medical Climatology, Physical Medicine Library*, Vol. 8, Elizabeth Licht, Publisher, New Haven, Conn., 753 pp.

Sargent, F., II, and Tromp, S. W. (Editors), (1964) *A Survey of Human Biometeorology*. World Meteorological Organization, Technical Note No. 65, Geneva, 113 pp.

Tromp, S. W. (1963) *Medical Biometeorology*. Elsevier Publishing Co., Amsterdam, 491 pp.

Tromp, S. W. and Bouma, J. J. (Editors) (1974) "Macro- and micro-environments in the atmosphere; Their effects on basic physiological mechanisms of man", *Progress in Biometeorology Div. A*, Vol. 1, Pts 1A & B, Swets & Zeitlinger, Amsterdam, 726 pp.

Tromp, S. W. and Bouma, J. J. (Editors) (1977) "Pathological Biometeorology," *Progress in Biometeorology Div. A*, Vol. 1 Pt. II, Swets & Zeitlinger, Amsterdam, 416 pp.

REFERENCES

Ambach, W. (1978) Kontaminierte Atmosphäre: Ist eine erhöhte Belastung des Menschen durch Anstieg der solaren UV-Strahlung zu erwarten? Zs. f. angew. Bäder- und Klimaheilkunde *25*, pp. 156-163.

Amelung, W. (1970) *Medizinische Klimatologie*, (Third edition) Dt. Bäderverband, Bonn, Drei Kronen-Druck & Verlag, Efferen, 31 pp.

Andersen, K. L., Hellström, B. and Eide, R. (1965); Strenuous Muscular Exertion in the Polar Climate and its Effect upon Human Work Capacity and Tolerance to Cold, *Final Report Contract AF EOAR 64-58*, US Air Force Office of Scientific Research, 30 pp.

Auliciems, A. and deFreitas, C. R. (1976) "Cold Stress in Canada, a Human Climate Classification," *Int'l. J. Biometeorol.* 20, pp. 287-294.

Aluiciems, A., and Kalma, J.D. (1979) A Climatic Classification of Human Thermal Stress in Australia.

Becker, F. (1972) Bioklimatische Reizstufen für eine Raumbeurteilung zur Erholung, Veröff Akad. f. Baumforschung und Landesplanung, Forschgs. und Sitzber. 76, pp. 45-61.

Becker, F. & Wagner, M. (1972) Die bioklimatischen Zonen der Bundesrepublik Deutschland, 1 sheet — Annex to Becker (1972).

Belisario, J. C. (1959) *Cancer of the Skin*. Butterworth & Co., Ltd., London.

Bormann, F. von (1960) Die klimatische Bedingtheit der epidemiologischklinischen Gesetzmässigkeiten einiger Seuchen, *Fundamenta Balneo-Bioclimatologica* 1, p. 335.

Brezowsky, H. (1964) "Morbidity and Weather," in S. Licht (Editor), *Medical Climatology*, Elizabeth Licht, Publisher, New Haven, Conn., pp. 358-399.

Brezowsky, H. and Menger, W. (1959) Vergleichende Untersuchungen über den Einfluss des Wetters in verschiedenen klimatischen Gebieten bei der meningitis epidemica, *Monatsschrift f. Kinderheilkinde,* no. 107, pp. 489-494.

Bristow, G. C. (1955) "How Cold is it?," *Weekly Weather and Crop Bull.* No. 42 (48), pp. 6-7.

Büettner, K. (1938) *Physikalische Bioklimatologie,* Akad. Verlagsgesellschaft, Leipzig, 155 pp.

de Freitas, R. C. (1979) "Human Climates of Northern China," *Atmospheric Environment.*

Edholm, O. G. (1966) "Problems of Acclimatisation in Man," *Weather* Vol. 21, pp. 340-350.

Falconer, R. (1968) Windchill, a Useful Wintertime Weather Variable, *Weatherwise* Vol. 21, pp. 227-229, 255.

Fanger, P. O. (1970) *Thermal Comfort.* Danish Technical Press, Copenhagen, 244 pp.

Federal Health Office (1965) *Short Climatic Guide to Switzerland.* Swiss Association of Climatic Resorts Vevey, 120 pp.

Flach, E. and Mörikofer, W. C. "Comprehensive Climatology of Cooling Power as Measured with the Davos Frigorimeter," *Physikal.-Meteorolog. Obs. Davos,* Contract Reports DA-591, Pt. I (1962), 60 pp; Pt. II (1965), 34 pp.; Pt. III (1966), 28 pp.; Pt. IV (1967), 30 pp.

Gilat, T., Shibolet, S. and Sohar, E. (1963) "The Mechanism of Heat Stroke," *J. Trop. Medicine and Hygiene* Vol. 66, pp. 204-212.

Gillet, J. D. (1974) "The Mosquito: Still Man's Worst Enemy," *Am. Scientist,* Vol. 61, pp. 430-436.

Gregorczuk, M. (1971) "Distribution of More Important Complex Biometeorological Indices on the Globe," *Ekologia Polska* Vol. 19, pp. 745-787.

Gregorczuk, M. (1971) "Bioclimatic Regions of the Globe with Special Consideration to the Area of Poland," *Ekologia Polska,* Vol. 19, p. 807-852.

Haase, C., and Leidreiter, W. (1975) "Stand der Klima- und Meteoropathologie," *Zeitschr. f. Meteorol.* Vol. 27, pp. 281-286.

Haenszel, W. (1962) "Variations in Skin Cancer Incidence within the United States," in F. Urbach (Editor), *Conference on Biology of Cutaneous Cancer, Ntl. Cancer Inst. Monogr.* No. 10, Washington, D. C., pp. 225-243.

Harlfinger, O. (1975) "Vergleichende Untersuchung der physiologischen Wärmebelastung zwischen Mitteleuropa und den Mittelmeerländern," *Arch. Met. Geophys. Biokl. Ser.* B, Vol. 23, pp. 81-98.

Hayes, R. D., and Hess, A. D. (1964) "Climatological Conditions Associated with Outbreaks of Eastern Encephalitis," *Am. J. Trop. Medicine and Hygiene,* Vol. 13, pp. 851-858.

Hollies, N. R. S. and Goldman, R. F. (Editors) (1977) *Clothing Comfort.* Ann Arbor Science Publishers, Inc., Ann Arbor, 189 pp.

Jäger, I. (1968) "Statistische Untersuchung über den zeitlichen Zusammenbang von Herzinfarkten und Apoplexien mit Grosswetterlagen und Wetterphasen," *Medizinische Welt* Vol. 19 (N.F.), pp. 1267-1275.

Jusatz, H. J. (1962) "The World Atlas of Epidemic Diseases and its Significance for

Bioclimatological Classifications," in S. W. Tromp (Editor) *Biometeorology.* (Proceed. 2nd Int'l. Biomet. Congr.); Pergamon Press, Oxford, pp. 141-146.

Jusatz, H. J. (1977) "Influence of Weather and Climate on the Geographical Distribution of Human Diseases," in S. W. Tromp and J. J. Bouma (Editors), *Progress in Biometeorology A,* Vol. I Pt. II, pp. 189-193, Swets and Zeitlinger, Amsterdam.

Kalma, J. D. and Auliciems, A. (1978) "Climate in Relation to Human Comfort," in E. G. Hallsworth and J. T. Woodcock, (Editors), *Land and Water Resources of Australia,* Proceed. 2nd Invitat. Sympos. Austral. Acad. Technol. Sci, pp. 279-299.

Kevan, S. M., and Faust, V. (1976) "Wetter-Klima-Kriminalität, *Zs. J. Allgemeine Medizin,* Vol. 52, pp. 257-264.

Ladell, W. S. S. (1957) "The Influence of Environment in Arid regions on the Biology of Man," *Human and Animal Ecology.* Arid Zone Research VIII, UNESCO, Paris, pp. 43-99.

Lawrence, J. S. (1977) "Influence of Weather and Climate on Rheumatic Disease," in S. W. Tromp and J. Bouma (Editors), *Progress in Biometeorology Div. A.,* Vol. 1, Pt. II, pp. 83-88, Swets and Zeitlinger, Amsterdam.

Lee, D. H. K., and Vaughan, I. A. (1964) "Temperature Equivalent of Solar Radiation on Man," *Int. J. Biometeorol,* Vol. 8, pp. 61-69.

Lipfert, F. W. (1980) "Sulfur Oxides, Particulates and Human Mortality: Synopsis of Statistical Correlations," *J .Air Pollut. Control Ass'n.* Vol. 30, pp. 366-371.

Loomis, W. F. (1967) "Skin-pigment Regulation of Vitamin-D Biosynthesis in Man." *Science,* Vol. 157, pp. 501-506.

Macfarlane, W. V. (1974) "Acclimatisation and Adaptation to Thermal Stress," in S. W. Tromp and J. J. Bouma (Editors), *Progress in Biometeorology Div. A,* Vol. 1, Pt. 1B, pp. 468-473.

Masterson, J. M. and Richardson, F. A. (1979) *"Humidex" a Method of Quantifying Human Discomfort due to Excessive Heat and Humidity.* Atmospheric Environment Service Downsview, Ontario, Canada, CLI 1-79, 43 pp.

May, J. M. (1950) "Medical Geography: Its Methods and Objectives," *Geogr. Rev.,* Vol. 40, pp. 9-41.

Mumford, A. M. (1979) "Problems of Estimating Lowland Windchill," *Weather,* Vol. 34, pp.424-429.

National Academy of Sciences (1979) *Stratospheric Ozone Depletion by Halocarbons.* Chemistry and Transport, Washington, D.C., 273 pp.

National Academy of Sciences (1975) "Estimates of Increases in Skin Cancer due to Ultraviolet Radiation Caused by Reducing Stratospheric Ozone," *Environmental Impact of Stratospheric Flight.* Washington, D.C., pp. 177-221.

National Academy of Sciences (1973) *Mosquito Control: Some Perspectives for Developing Countries.* Washington, D.C., 63 pp.

Quisenberry, W. B. (1962); Ethnic Differences in Skin Cancer in Hawaii, in F. Urbach (Editor), *Conference on Biology of Cutaneous Cancer, Ntl. Cancer Inst. Monogr.,* No. 10, Washington, D. C., pp. 181-189.

Robinson, N. (1966) *Solar Radiation.* Elsevier Publishing Co., Amsterdam, 347 pp.

Schmidt-Kessen, W. (1977) "Therapy in Natural Climates," in S. W. Tromp and J. J. Bouma (Editors), *Progress in Biometeorology Div. A,* Vol. 1, Pt. II, pp. 258-261, Swets and Zeitlinger, Amsterdam.

Segi, M. (1962) World Incidence and Distribution of Skin Cancer, in F. Urbach (Editor), *Conference on Biology of Cutaneous Cancer, Ntl. Cancer Inst. Monogr.* No. 10, Washington, D. C., pp. 245-255.

Siple, P. A. and Passel, C. F. (1945) Measurement of Dry Atmospheric Cooling in Subfreezing Temperatures, *Proc. Am. Phil. Soc.,* Vol. 89, pp. 177-199.

Spangenberg, W. W. (1954) Bemerkungen zur Therapeutischen Wirkung des Bettes, D. Deut. Gesundheitswesen, Vol. 9, pp. 1057-1061.

Stamp, L. D. (1964) *Some Aspects of Medical Geography.* Oxford University Press, London, 103 pp.

Steadman, R. G. (1971) Indices of Wind Chill of Clothed Persons, *J. Appl. Meteorol.,* Vol. 10, pp. 674-683.

Sulman, F. G. (1976) *Health, Weather and Climate.* S. Karger, Basel, 166 pp.

Sulman, F. D., Levy, D., Levy, A., Pfeifer, Y., Superstine, E. and Tall, E. (1974) "Air-ionometry of Hot, Dry Des(s)ert Winds (Sharav) and Treatment with Air Ions of Weather-Sensitive Subjects," *Int'l. J. Biometeorol.,* Vol. 18, pp. 313-318.

Swantes, H. J. (1971) *Probleme der Medizin-Meteorologie im Zusammenhang mit einer Untersuchung von O. Höflich über die Wetterabhängigkeit von Phantomschmerzen.* Beilage zur Berliner Wetterkarte 113/71 — SO 34/71, 9 pp.

Swantes, H. J. and Reinke, R. (1978) "Föhn-Wetter-Mensch," *Dt. Ärzteblatt-Ärztliche Mittlg.,* Vol. 71, 1789-1790.

Tal, E., & Sulman, F. G. (1973) "Thyroid Reaction in Heat Stress," *Isr. Pharmaceut. J.,* Vol. 16, p. 82.

Terjung, W. H. (1966) "Physiologic Climates of the Conterminous United States: A bioclimatic Classification based on man," *Annals, Ass'n. Am. Geogr.,* Vol. 56, pp. 141-179.

Terjung, W. H. and Louie, S. S-E. (1971) "Potential Solar Radiation Climates of Man," Annals Ass'n. *Am. Geogr.,* Vol. 61, pp. 481-500.

Tromp, S. W., Faust, V. (1977) "Influence of Weather and Climate on Mental Processes in General and Mental Diseases in Particular," in S. W. Tromp (Editor), *Progress in Biometeorology 1,* (Pt. II), Swets and Zeitlinger, Amsterdam, pp. 74-82.

Wehner, A. P. (1969) "Electro-aerosol, Air Ions and Physical Medicine," *Am. J. Phys. Medicine,* Vol. 48, pp. 119-149.

CHAPTER THREE

Climate and Agriculture

M. S. Swaminathan
International Rice Research Institute, Manila

INTRODUCTION

THE EARLIEST HUMAN form is dated some 25 million or more years ago. From the early hominids such as *Zinjanthropus,* it took many million years more for Homo sapiens to develop. In spite of the antiquity of the evolutionary processes resulting in the birth of modern man, the art of cultivating crops was developed only about 12,000 years ago. Until the era of settled cultivation, man met his calorific and other nutritional needs by gathering food from natural vegetation and by hunting and fishing. The transition from food gathering to food growing was marked by several important changes in human life, because the opportunities afforded by settled agriculture cradled culture.

Settled cultivation brought in its wake many new problems, such as the erosion of soil fertility, the dependence of crop yields on weather behaviour, and the incidence of pests and diseases. Through trial and error, early cultivators found answers to these problems. Shifting cultivation provided a means of restoring soil fertility. Total failure of crops due to moisture stress or moisture excess was avoided to some extent through mixed cropping and inter-cropping. Damage by pests and diseases was minimised by growing a mosaic of crop varieties, with considerable genetic variability in the material.

The birth of agriculture also gave rise to the introduction of various forms of energy, collectively referred to as cultural energy (Figure 3.1). Irrigation, an important component of the cultural energy forms introduced into agriculture, has been an important factor in elevating and stabilising crop production. In spite of the steps taken so far, it has not been possible to insulate agricultural fortunes from the vagaries of the weather.

The very process of change in the relationship between man and his environment initiated by agriculture has had several adverse effects in terms of long term production prospects. Shifting cultivation, soil erosion, indiscriminate deforestation, and other human activities, such as mining for minerals, have resulted in either the destruction or diminution of the biological potential of land in nearly one-third of the

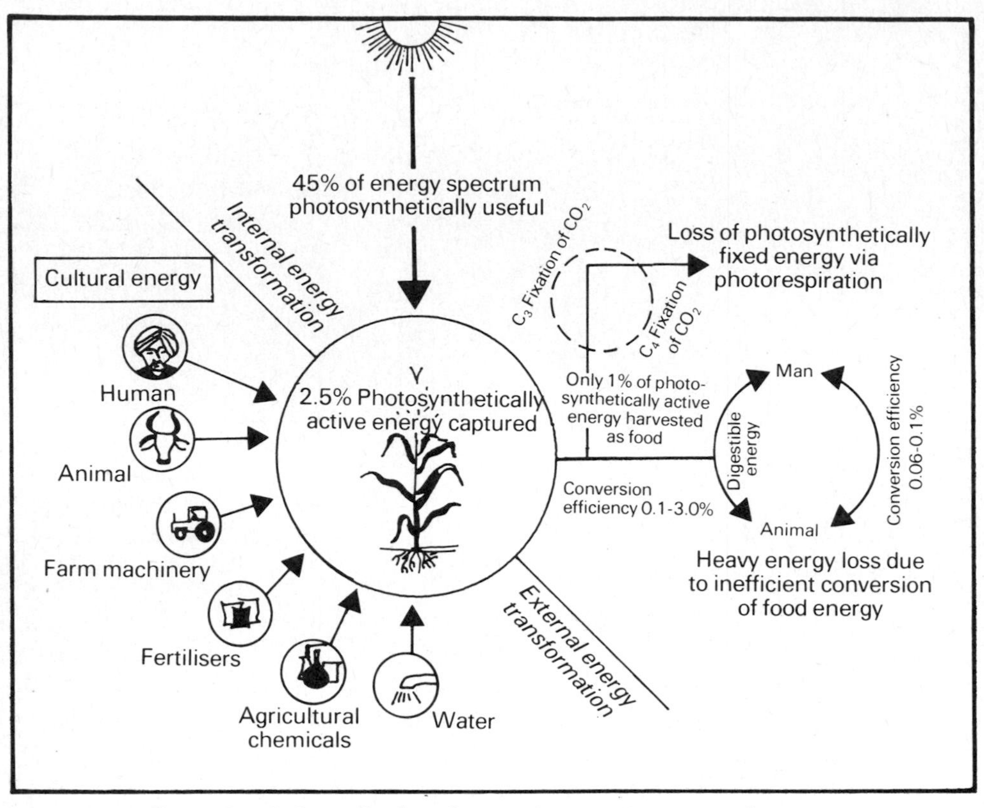

Figure 3.1. Solar and cultural energy input and output cycles in rice

world's surface. Today, several of the early centres of domestication are characterised by a high degree of desertification. Even in parts of the fertile crescent area in the Middle East, where the early domestication of crops occurred, desert conditions now exist.

The challenge lies in developing methods of sustainable agricultural growth. This is where a detailed study of the relationships between climate and agriculture assumes importance.

VULNERABILITY OF WORLD FOOD PRODUCTION SYSTEMS

Rapid increase in population, dependence on too few crops and animals for meeting global food needs, and dependence on too few countries for balancing the global food budget have led to a situation in which the conquest of hunger has become the most difficult as well as the most urgent task (Figures 3.2 to 3.5). Weather-induced

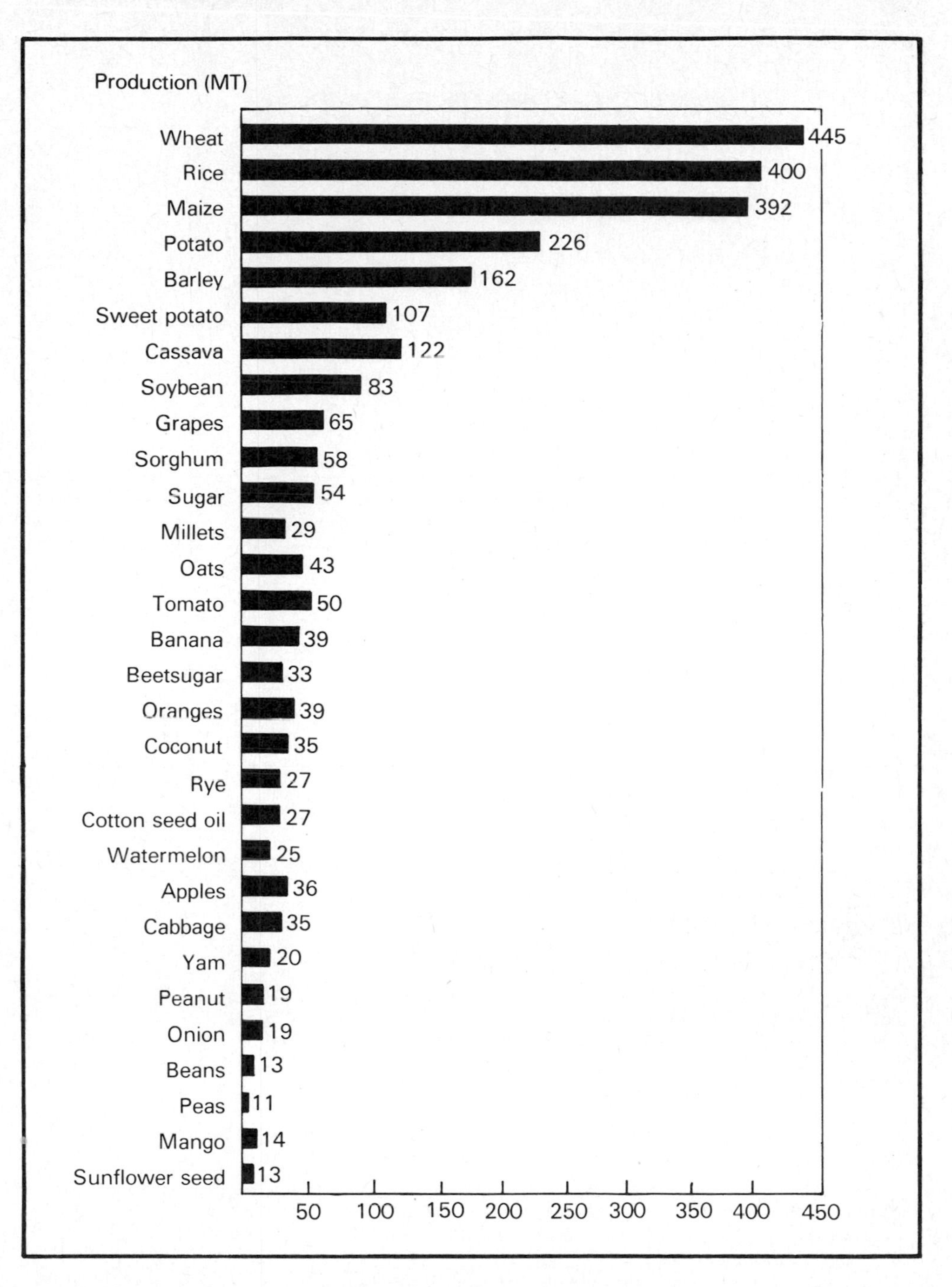

Figure 3.2. Annual production of the world's major food crops 1980 (SOURCE: FAO Production Yearbook, 1980)

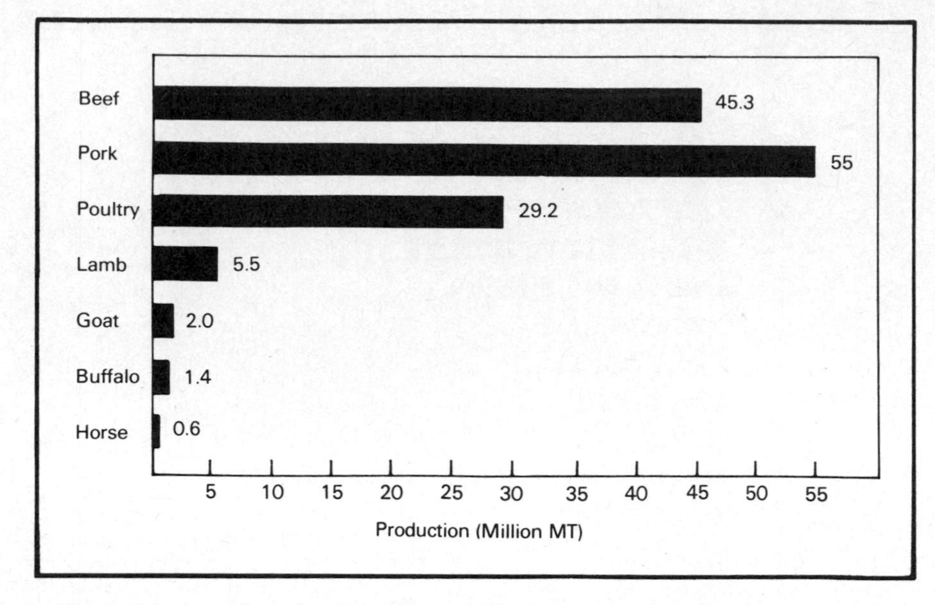

Figure 3.3. Annual production of the world's major animal products (SOURCE: FAO Production Yearbook, 1980)

Figure 3.4. World's increasing dependence on the grain exports of a few countries; USA and Canada supply most of the grain

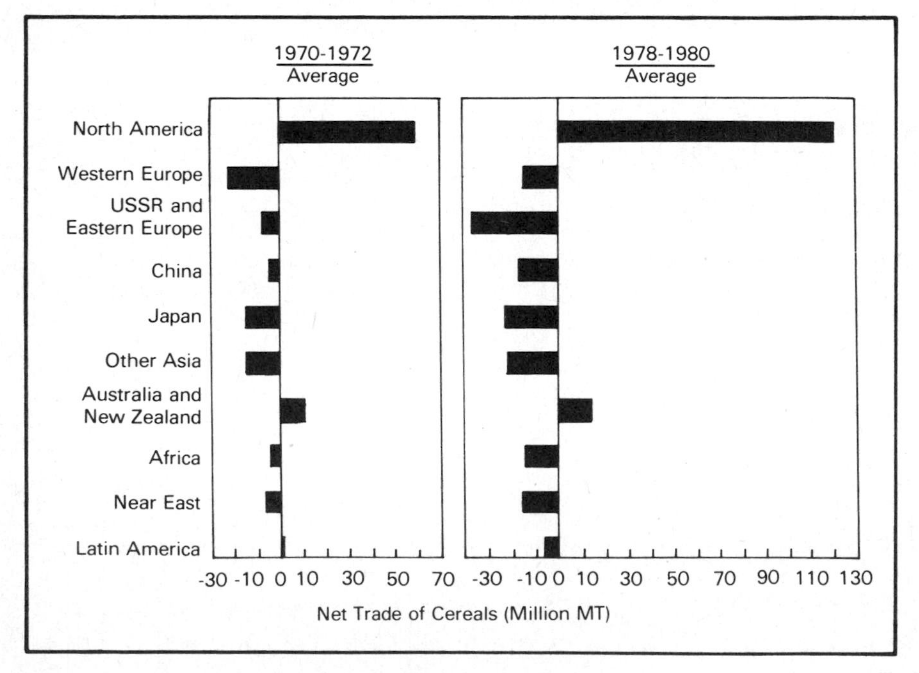

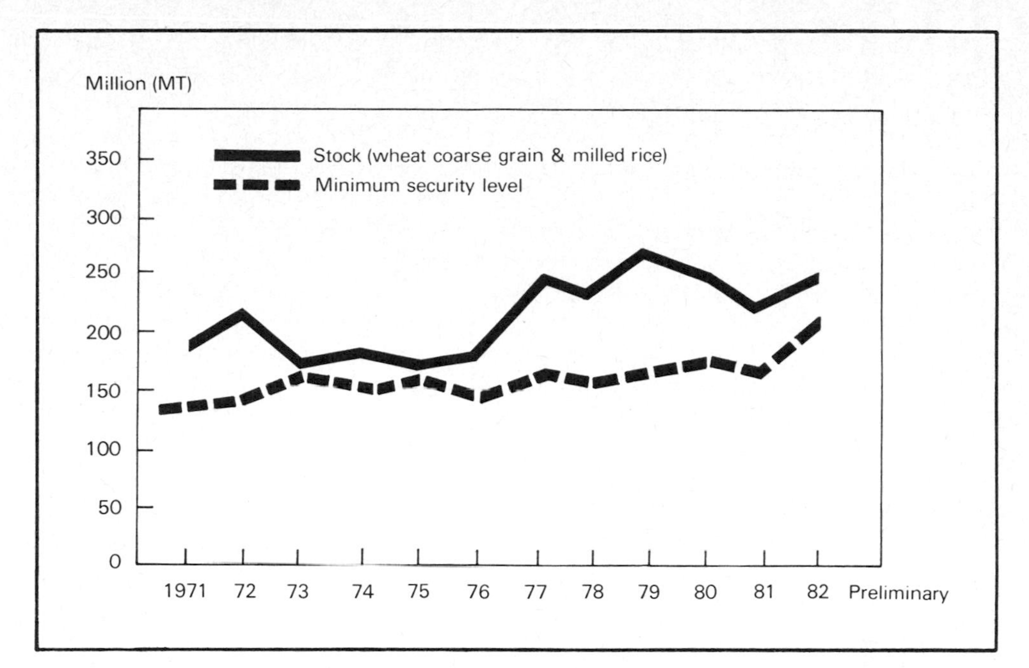

Figure 3.5. World carry-over stocks and minimum security levels of cereals (including China and USSR

fluctuations in production have led expectations for global food security to oscillate between hope and despair. A 5 to 10 per cent increase or decrease in production results in either an uncomfortable glut or an acute scarcity. Consequently, public policies in the matter of insurance, pricing, marketing and distribution of farm inputs and outputs have become important determinants in the control of production fluctuations. Weather, pest epidemics and public policies have important effects on stability of production.

According to FAO, world cereal production in 1981 exceeded expected global utilisation for the first time in three years, leading to a rise in carryover stocks at the end of the 1981-82 season and a definite improvement in the global food security situation.

The improvement in the overall supply situation was due to record crops in the main wheat and coarse grain exporting countries as well as in the major rice importing and exporting countries. This expansion reflected partly the higher prices prevailing in 1980 and partly the favourable growing conditions. Production of non-cereal foodstuffs also rose in developing countries. A rise in world cereal production of 7 per cent in 1981 exceeded the trend for the first time since 1978, despite lower harvests in Europe and the USSR. Output in the developing countries rose by 6 per cent, well above the average growth of the last decade.

As a result, global cereal stocks, which were drawn down heavily during 1979-80 and 1980-81, are provisionally forecast to rise to 270 million tonnes at the end of the

1981-82 season, representing nearly 18 per cent of world consumption. International prices of wheat and coarse grains have been substantially lower during the 1981-82 season. Prices of rice have declined by almost 40 per cent since mid-1981.

While there is an improvement in the overall global situation, the fact remains that the food import needs of many developing countries have increased dramatically in recent years. The cereal imports of developing countries as a whole have doubled in the past decade and are now close to 100 million tonnes a year.

Of considerable concern is the situation that had developed in low-income countries. According to an analysis made by FAO, as many as 37 low-income countries (with per capita incomes of US$730 or less in 1980) recorded negative growth rates in per capita production of cereals during the seventies; of these, 19 experienced a decline in total cereal output. Thus, far from achieving greater self-sufficiency, these countries are faced with a widening food gap. Food aid received by these countres has remained stagnant in recent years. The low-income countries as a whole now spend over US$7 billion a year to cover the value of commercial imports of cereals.

Food security in the world remains uncertain in the medium term. While international market prices of cereals have declined in 1982, following a rise for some years, costs of production have continued to rise. This has led to a concern in exporting countries that farmers' incentives will weaken and that this may have implications for growth in 1982 and even beyond. The US government limited the 1982 production of wheat, coarse grains, and rice through voluntary set aside programmes. According to the USDA, the initial projections of 1982-83 crops in the United States point to a decline of nearly 4 million tonnes of wheat and 16 million tonnes of coarse grains from 1981-82 production.

NATURAL RISKS IN AGRICULTURE

Agriculture is probably one of the riskiest professions in the world. Natural hazards affecting farm output include uncertainties of weather, such as deficiency of moisture or drought; excessive moisture, including flooding; excessive cold; hail; typhoons, tornados, cyclones, or windstorms; and natural fire and lightning.

These risks may be called meteorological risks. Schultz (1953), in a detailed analysis of the nature and factors of yield instability in the United States, stated with particular reference to west central regions, both north and south: "In this large area the hand of nature lifts and depresses yields despite all the efforts of farmers to counteract its influence." For example, in the drought year of 1936, wheat production in the United States fell to 526 million bushels, while production in the normal year of 1931 was 941 million bushels. Wide fluctuations in USSR food grain output are largely due to weather factors. Consequently, the USSR may import as much as 40 million tonnes of cereals in 1981-82, against 15 million tonnes three years ago.

The United Kingdom is considered to be relatively safe from exposure to severe weather hazards. But important exceptions occur now and then. For example, in 1976, alternate storms and drought dominated the agricultural scene. For many farmers, they caused havoc to buildings, equipment, crops, and production plans.

Precise estimates of the crop losses caused by weather aberrations are virtually impossible because of the difficulties in separating the impact of components such as management, technology, and climate (Biswas, 1980). Nevertheless, crop-climate models can be constructed if adequate data are available. Biswas (1980) has cited the possibilities in this area. The establishment of crop-weather watch groups on the lines recommended by Swaminathan (1982) in the different agro-ecological regions of every country would help in studying the crop-weather interrelationships. Ray (1981) has described the difficulties currently experienced in the field of agricultural insurance due to gaps in the data base essential for assessing the insurability of agricultural risks.

In spite of these difficulties, it is clear that some regions, countries, and parts of countries are more prone to climatic variables than others. In Africa, apart from the Sahelian countries which face chronic food problems, many countries have experienced wide fluctuations in production arising largely from the vagaries of the weather in recent years. In the Philippines, the yearly losses in rice production caused by pests, drought, typhoons, and floods have been worked out by Pantastico and Cardenas (1980). Their data are reproduced in Figure 3.6. Stansel (1980) has worked out the likely deviations from average rice productions caused by changes in precipitation in the United States (Table 3.1). Temperature at the grain filling stage is an important determinant of total yield.

Figure 3.6. Yearly losses in Philippine rice production, 1968-77

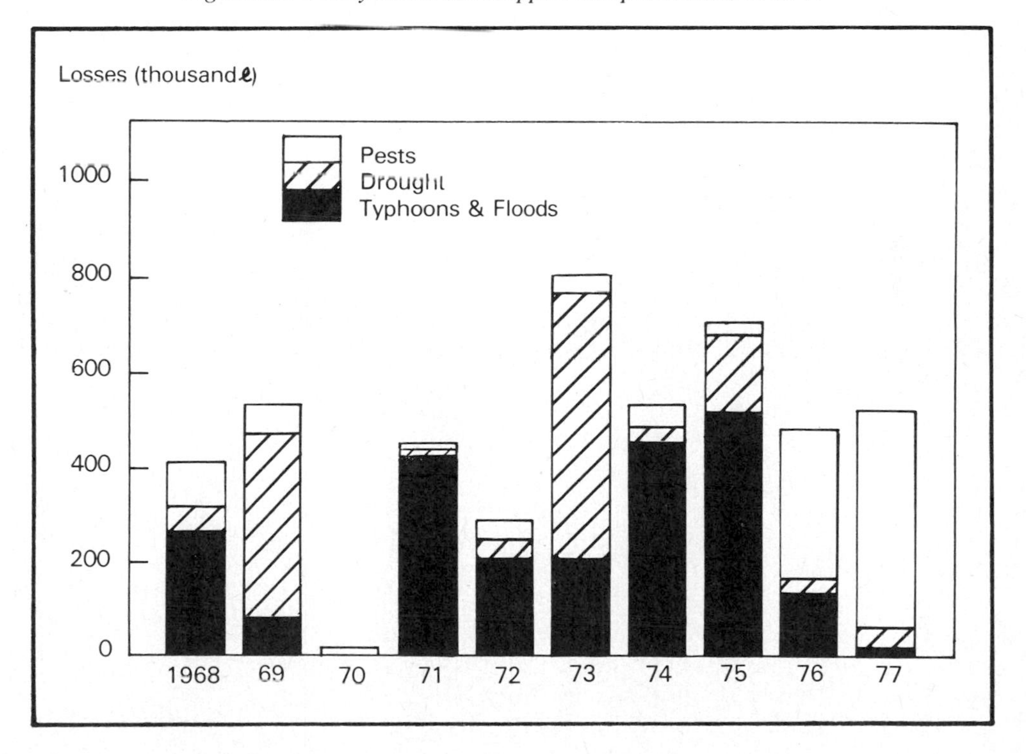

Table 3.1. Deviation from US average production of rice as influenced by changes in precipitation and temperature[1]

Precipitation change (%)	Deviation (%) at a temperature change of -2°C	-1°C	-.5°	0	+.5°C	+1°C	+2°C
-15	2	5	8	8	11	15	20
-10	-1	2	5	6	8	12	18
-5	-4	-1	2	2	5	9	14
0	-9	-4	-1	0	4	7	12
+5	-9	-6	-3	-2	0	4	9
+10	-14	-10	-7	-6	-3	0	6
+15	-18	-14	10	-9	-6	-2	3

1. Based on US production of 5 million t.

IMPACT OF DROUGHT ON AGRICULTURE

Drought, floods, cyclones, typhoons, hailstorms, frost, dew and snowfall all influence crop production to varying degrees. Of all these phenomena, widespread drought has historically had the most detrimental effect. Therefore, it would be useful to discuss the relationships between drought and agriculture in some detail.

Agricultural scientists and meteorologists tend to adopt different definitions of drought, the former in terms of crop behaviour and the latter in terms of rainfall. If, for example, the total rainfall over India (112 cm yr^{-1}) was evenly distributed in space and time, the water needs for the entire cultivated areas would be satisfied.

Agricultural droughts occur in particular areas because of erratic rainfall unrelated to crop needs. At a symposium on Meteorological Aspects of Tropical Droughts held in December 1981 at New Delhi under the sponsorship of the World Meteorological Organization, these conclusions on the relationship between drought and agriculture were drawn:

(a) The fundamental requirement for drought management in India is the collection and storage of water for utilisation in protective irrigation during dry spells (water harvesting).

(b) Food-grain storage measures taken to combat drought were extremely successful in 1979, the most severe meteorological drought this century. The result was no starvation and no need to import.

(c) Agricultural yield correlates poorly with annual rainfall, but well with a crop index based on crop need and soil water content. The latter may be included in the definition of the climatology of a particular region as a soil moisture index and used for agricultural drought evaluation.

(d) Refined predictions of rainfall are needed to match crops with expected weather, particularly now that so many hybrid varieties of crops are available.

The symposium participants recommended:

(a) Good forecasts of rainfall, both of quantity and distribution, are needed for agriculture in the short, medium, and long term, particularly during the monsoon season. Both probabilistic and dynamic forecasting methods should be further developed. Probabilistic forecasts should be issued until deterministic forecasts are demonstrably superior.

(b) The dialogue between meteorologists and agriculturalists and soil scientists must continue and FAO and UNDP should be encouraged to support this activity, e.g. by setting up special training programmes on meteorological applications for agriculturalists.

(c) Better communications and exchange of information in this area should be set up between developed (high technology) and developing countries, e.g. through exchange visits and symposia.

An integrated strategy to minimise the adverse impact of drought on crops, farm animals, and human population with short and long term action plans will have to be developed at the national level. In the long term strategy, the development of irrigation sources and scientific land and water use planning will have to receive attention. Also, the capability for making fairly accurate long term weather forecasts will have to be developed.

Fortunately, the various coordinated programmes of research sponsored by WMO, FAO, UNESCO, and national and international organisations are now providing valuable data. For example, Oldeman and Frère (1982) have provided detailed information on the agroclimatology of the humid tropics of Southeast Asia. Detailed data also are available for many other regions of the world, including the Sahel Region of Africa which underwent prolonged drought during the seventies. Several scientists such as Campbell and Bryson (1982) have been publishing year-in-advance forecasts of monsoon rainfall in India, USA, and other countries. Studies on crop yield and climate up to the year 2000 have been published by institutions in the USA as a part of their research on climate impact assessment (National Defense University and USDA, 1980).

Mooley *et al.* (1981) have compiled available information on the variability of annual rainwater availability in India. Their analysis reveals that water deficiency has had a much greater impact on agricultural production than water excess (Tables 3.2 and 3.3). Hence, developing countries have wisely stepped up their investments in irrigation (Table 3.4). This is particularly important because the only pathways open to many developing countries, particularly in Southeast Asia, for producing more food grains to meet the needs of the expanding population are higher productivity per unit area, water, and other cultural energy inputs and higher intensity of cropping through multiple and relay-cropping. This is because of the small land area available per capita population in Asia (Table 3.5). Arnon (1981) has recently discussed the resources, potentials, and problems in the modernisation of agriculture in developing countries.

An example of how crop production can be improved and stabilised through investment in irrigation is provided by the trends in crop production in India. The area under irrigation has been steadily improving (Table 3.6); hence the trough point in favourable years is also better (Table 3.7). Even during 1979, the most severe meteorological drought year of this century, food production was 30 million tonnes more than during the very favourable monsoon year of 1964-65 (Table 3.7). Swaminathan (1982) has shown how in areas such as the Punjab, where the pumping

Table 3.2. Percentage change in the yearly All India Index of food grain production in the year of marked annual rain-water deficiency over India

Year of marked deficiency	Percentage change in the index in the year
1904	–2.1
1905	–4.9
1911	–4.6
1918	–32.3
1920	–24.1
1941	4.5
1951	1.1
1965	–19.6
1966	1.7
1972	–8.2
1974	–5.4

Note: Values of the index of agricultural production are not available prior to 1900. Since 1965, 1966 are consecutive years of marked deficiency, there is little change in the Index of Agricultural Production in 1966.

Table 3.3. Years of marked rain-water deficiency over India (1871-1978)

Year	Percentage of the country's area under rain-water deficiency	Total annual rain-water deficiency over area under deficiency in km^3
1877	40.2	332.36
1899	75.1	901.21
1904	44.9	321.91
1905	47.1	401.70
1911	41.2	341.10
1918	60.0	649.18
1920	46.1	416.67
1941	55.0	432.45
1951	44.8	298.53
1965	52.0	447.65
1966	41.2	381.24
1972	63.5	571.93
1974	41.3	377.83

Table 3.4. Increase in areas under irrigation (million hectares) during the period 1961-65 (annual average to 1977 (FAO, 1978)

	1961-65	1977	Increase (%)
Africa	5.8	7.8	13.4
Asia	100.0	128.8	28.8
South America	4.9	6.5	32.6
All developing countries	110.0	144.9	31.7
All developed countries	39.0	52.9	35.6
World	149.0	198.0	32.9

Table 3.5. Arable land per head of population (calculated from data from FAO, 1978)

	Per capita agricultural population (ha)	Per capita total population (ha)
Africa	0.72	0.48
Asia	0.32	0.19
South America	1.40	0.46
All developing countries	0.42	0.26
All developed countries	4.39	0.59

of groundwater is the most important source of irrigation, energy management holds the key to reducing a drop in food output during a drought year. With effective energy management, the production of rice in the Punjab during the severe drought year of 1979-80 was maintained at about the same level as in 1978-79. Rainfall in 1978-79 was normal (Table 3.8). The Punjab has over 80 per cent of its cultivated area under irrigation, Madhya Pradesh and Orissa have less than 30 per cent.

Research on drought management in India

Drought is a recurrent phenomenon in the dry areas of India, as it is in all of the semi-arid and sub-humid areas of the world. In these regions, annual loss of water through evaporation and transpiration is greater than annual precipitation. In addition, annual rainfall in the semi-arid tropical areas usually is distributed unevenly. For instance, Hyderabad, Andhra Pradesh, receives over 80 per cent of its 780 mm rainfall between June and October. In Jodhpur, Rajasthan, almost all the annual precipitation (350 mm) falls between July and September.

With low precipitation relative to evaporation and transpiration, a short rainy season with uneven seasonal distribution of rainfall, a great interannual variability and small negative deviations in precipitation are all that is required to initiate drought. Small and large negative deviations are highly probable and even certain to

Table 3.6. Peak and trough points in foodgrain production during 1960-61 to 1978-79 (million tonnes)

Year	Foodgrains production. Adjusted actual production	Peak point	Trough point
1960-61	82.3		82.3
1961-62	82.4	82.4	
1962-63	80.3		80.3
1963-64	80.7		
1964-65	89.4	89.4	
1965-66	72.3		72.3
1966-67	74.2		
1967-68	95.1	95.1	
1968-69	94.0		94.0
1969-70	99.5		
1970-71	108.4	108.4	
1971-72	105.2		
1972-73	97.0		97.0
1973-74	104.7	104.7	
1974-75	99.8		99.8
1975-76	121.0	121.0	
1976-77	111.2		111.2
1977-78	126.4		
1978-79	131.4	131.4	
1979-80	109.0		109.0

Table 3.7. Variability in rice production in four states of India (Production in lakh tonnes)

State	1977-78	1978-79	1979-80	1980-81
Madhya Pradesh	44.4	35.6	18.3	40.0
Orissa	43.2	44.0	29.2	43.3
Punjab	24.9	30.9	30.4	32.2
Uttar Pradesh	52.0	59.6	25.5	54.4

occur in such regions. A survey reported by Ryan (1974) revealed that moderate or worse droughts are likely to occur in semi-arid India one year in every four.

There have been many definitions of drought. A simple definition of agricultural drought is given by Rosenberg (1980). Drought is a climatic excursion involving a shortage of precipitation sufficient to adversely affect crop production or grassland or horticultural productivity. The traditional farmer in the rainfed semi-arid tropics lives with the possibility of drought. He has learned by experience how to adjust to the inherent variability of climate. Each year he must decide what to plant, where and

Table 3.8. Disease and insect resistance reactions of varieties named by IRRI and named by the Philippine government

	Disease and Insect Reactions[1]							
	Blast	Bacterial blight	Grassy stunt	Tungro	Green Leaf-hopper	Brown Plant-hopper	Stem-borer	Gall midge
IR5	MR	S	S	S	R	S	MS	S
IR8	S	S	S	S	R	S	S	S
IR20	MR	R	S	MR	R	S	MR	S
IR22	S	R	S	S	S	S	S	S
IR24	S	S	S	S	R	S	S	S
IR26	MR	R	MR	MR	R	R	MR	S
IR28	R	R	R	R	R	R	MR	S
IR29	R	R	R	R	R	R	MR	S
IR30	MS	R	R	MR	R	R	MR	S
IR32	MR	R	R	MR	R	R	MR	R
IR34[2]	R	R	R	R	R	R	MR	S
IR36[2]	R	R	R	R	R	R	MR	R
IR38[2]	R	R	R	R	R	R	MR	R
IR40[2]	R	R	R	R	R	R	MR	R
IR42[2]	R	R	R	R	R	R	MR	R

1. S - Susceptible; MS - Moderately susceptible; MR - Moderately resistant; R - Resistant. Reactions based on tests conducted in the Philippines for all diseases and insects except gall midge. Screening for gall midge was done in India.
2. Named by the Philippine Government.

when. He will consider mainly what will survive until harvest to meet the needs of his family.

However, the Indian economy is largely dependent on agriculture. Experience indicates that agricultural drought can have ramifications to all sectors of the economy and society. The effects of drought on agricultural production in the Indian sub-continent are reasonably well understood — certainly in a qualitative sense. We also know the kinds and extent of the multiple effects of agricultural drought on the regional and national economies and the national and interregional food trade. The effects of drought on the social and political conditions of the affected regions and the nations are somewhat understood. Scattered information on the direct and indirect impacts of drought on urban life and urban society, and on the nutritional status of various segments of society is available.

Many specific technological drought-mitigating measures have been proposed, such as water storage and recycling, rainfall enhancement, evaporation suppression, alternative agronomic management techniques, proper selection of drought tolerant crops and their cultivars, and use of wind breaks. However, a thorough study and quantitative evaluation of the potential and practicality of these measures in regular and emergency use have not yet been attempted on any large scale. Similarly, it can be said that the efficacy of economic measures proposed to moderate drought effects, e.g. cost-sharing programmes, crop insurance, credit supports, and cooperative movements, also have not yet been fully evaluated.

There are examples of national planning for drought contingency or for response to drought once it occurs. In 1979, a serious drought developed in nearly half of the Indian sub-continent due to the failure of the south-west monsoon. The Government of India instituted a major programme of relief, major elements of which were described by Swaminathan (1979 and 1982).

Some of the components of this programme were: first, a *Food for Work* scheme, where farmers and others lacking work because of crop failure were employed by the government in the building of durable structures in rural areas for increasing agricultural productivity in normal rainfall years. Preference is given to construction of small irrigation structures, which should help provide relief in subsequent droughts; second, a *Food for Nutrition* programme to provide food to those who are unable to work, such as the aged and infirm, pregnant and lactating mothers, and children; third, a provision of *Water Security* of drinking water for man and animals; fourth, an especially innovative idea, the designation of certain districts as *Most Favourable Areas* which were more favoured in recent rainfall, and in which the greatest productivity of a post-rainy season crop can be achieved by diverting to them available resources of fertilisers, good quality seed and other requisites. By stepping up relief measures in the most seriously affected regions and by intensifying production effects in the most favourable areas, hardship to human and animal populations can be minimised. This provides one excellent example of a consistent national strategy.

Past approaches to resource development for increasing agricultural production in the rainfed semi-arid tropics have achieved very limited success because they have not recognised the basic climatological and soil characteristics of the region nor utilised natural watershed and drainage systems (Kampen and Burford, 1980). Better technologies are now being developed by the All India Coordinated Research Project for Dryland Agriculture (AICRPDA), State Agricultural Universities, and the International Crops Research Institute for the Semi-Arid Tropics (ICRISAT) to ameliorate the effects of drought, to increase food production per unit of land and capital, to assure stability, and to contribute directly to improving the quality of life. Some of the components of this technology are:

— improved soil management practices;
— use of improved and appropriate seeds;
— use of fertilisers;
— intercropping/double cropping;
— proportionate cropping;
— watershed management;
— supplementary life saving irrigation.

The researches conducted by AICRPDA and ICRISAT over the past decade or so at their research centres and in operational scale, village-level studies have shown that a technology for land and water management for a vast area of India (the black soil region) can substantially contribute to increased and stable crop production. Murthy (1981) has estimated that the area under such soils in India is 72.0 million hectares, which is about 22.2 per cent of the geographical area of the country. The area falls under two rainfall classes: dependable, in which stable dryland agriculture is feasible, and undependable, where risks to crop-based dry farming are high. Such areas are suited for range farming and silvi-pastoral systems.

The watershed-based management system for vertisols consists of bringing together several components of improved technology and recognises that the improvement of any one component may have small effect on crop yields. But the combination of all the components produces spectacular results. For any cropping system, the three components that have produced the most significant synergistic effect are:

- land management practices that reduce runoff and erosion and that give improved surface drainage with better aeration and workability of the soils;
- cropping systems and crop management practices that establish a crop at the very beginning of the rainy season, that make efficient use of moisture throughout both the rainy and post-rainy seasons, and that give high sustained levels of yields;
- implements for cultivation, seeding, and fertilising that enable the required land and crop management practices to be efficiently carried out.

The data from operational-scale field experiments on vertisols at ICRISAT Centre for a maize/pigeon-pea intercropping system under rainfed conditions show that the highest yields (2610 kg ha^{-1}) were obtained by a combination of all these factors. The effect became more marked when fertiliser, soil and crop management system, and high-yielding variety were combined (3.5 t ha^{-1}). Thus, for improving crop production from vertisols under rainfed conditions, improvement of soil fertility and soil management are crucial. There is considerable evidence that the situation under irrigated farming is similar (Kanwar *et al.*, 1981).

Action plan for drought management

Enough expertise should be developed to classify the regions affected by adverse weather into broad categories:

Most seriously affected areas (MSA): in these areas, the priority concern should be to provide adequate relief and rehabilitation measures to avoid distress to human and animal populations. The provision of safe drinking water in addition to emergency food and nutrition programmes will need attention.

Most favourable areas (MFA): in these areas, either assured irrigation facilities exist or adequate soil moisture is available because of normal rainfall. MFAs could receive added attention by initiating compensatory production programmes.

Briefly stated, a national agricultural strategy in a year characterised by widespread drought will have three major components:

- popularisation of crop-saving measures and risk distribution agronomy;
- alternative cropping strategies to suit different weather probabilities and mid-season corrections in crop planning;
- compensatory production programmes in areas with adequate soil moisture (MFA).

To implement that strategy, it will be necessary to build reserves of seeds of alternate crops and of fertiliser. Just as grain reserves are important for food security, seed and fertiliser reserves are necessary for the security of crop production.

Fortunately, considerable research has been done in recent years at the International Crops Research Institute for the Semi-Arid Tropics and All India Coordinated Research Project on Dryland Agriculture. Some examples of results of the All India Coordinated Project are:

Crop substitution: After identifying the environment in terms of both land and water (rainfall), these crops were found to be more efficient than the traditional crops:

Region	*Traditional crop*	*Substitute crop*
Deccan region	Cotton	Safflower
	Wheat	Safflower
Malwa plateau	Wheat	Chickpea
		Safflower
Eastern Uttar Pradesh	Wheat	Mustard
Uplands of Chotanagpur and Orissa region	Rice	Groundnut
		Greengram
		Ragi *(Eleucine corocana)*
Black soils of South-East Rajasthan	Maize	Sorghum

Such data help in introducing mid-season correction in crop scheduling.

Intercropping: Promising intercropping systems for various regions are:

Centre	*System*	*Ratio*
	Sorghum-based	
Hyderabad	Sorghum + pigeon pea	2:1
Akola	Sorghum + mung bean	1:1
	Sorghum + blackgram	1:1
	Sorghum + pigeon pea	2:1
Indore	Sorghum + pigeon pea	2:1
Udaipur	Sorghum + soyabean	2:2
	Sorghum + pigeon pea	1:1
	Pearl millet-based	
Solapur	Pearl millet + pigeon pea	2:1
Bijapur	Pearl millet + pigeon pea	2:1
Rajkot	Pearl millet + pigeon pea	2:1
	Maize-based	
Udaipur	Maize + pigeon pea	1:1
Indore	Maize + soyabean	1:1
Dehra Dun	Maize + soyabean	8:2
Bhubaneswar	Maize + horsegram	1:1
Ranchi	Maize + pigeon pea	2:1
Rakh Dhiansar	Maize + greengram	1:1

Centre	*System*	*Ratio*
	Rice-based	
Bhubaneswar	Rice + pigeon pea	4:2
Ranchi	Rice + pigeon pea	4:2
	Finger millet-based	
Bangalore	Finger millet + leucerne (as fodder)	3:1
Bhubaneswar	Finger millet + horsegram	2:1
	Peanut-based	
Anantapur	Peanut + pigeon pea	5:1
	Peanut + castor	5:1
Bangalore	Peanut + pigeon pea	4:1
Rajkot	Peanut + pigeon pea	6:1
	Peanut + castor	6:1
	Pigeon pea-based	
Varanasi	Pigeon pea + blackgram	1:1
Bijapur	Pigeon pea + setaria	2:1
Solapur	Pigeon pea + setaria	2:1
	Castor-based	
Jodhpur	Castor + cowpea (fodder)	1:1
Dantiwada	Castor + cowpea	1:1
	Castor + sorghum	1:2
Hyderabad	Castor + sorghum	1:1
	Fodder-based	
Jodhpur	Cenchrus + cluster bean (fodder)	2:2
	Cenchrus + cluster bean (grain)	2:2
Anand	Dicanthium + cluster bean/siratro	As a
	Cowpea(fodder)	mixture

Promising cropping sequences are:

1st crop	*2nd crop*	*Centre(s)*
	Cereal-based sequence	
Rice	Wheat	Dehra Dun, Rewa
	Gram	Dehra Dun, Rewa, Varanasi
	Rice	Bhubaneswar
	Linseed	Ranchi
Maize	Wheat + gram	Hoshiarpur, Rakh Dhiansar
	Barley	Dehra Dun

1st crop	*2nd crop*	*Centre(s)*
	Cereal-based sequence	
	Gram	Dehra Dun, Hoshiarpur, Varanasi
	Mustard/rape seed	Hoshiarpur, Rakh Dhiansar
	Safflower	Ranchi, Udaipur, Indore
	Linseed	Ranchi
Sorghum	Safflower	Indore, Akela, Udaipur
	Gram	Indore, Rewa
Pearl millet	Mustard/rape seed	Agra, Rakh Dhiansar
	Gram	Varanasi
	Safflower	Varanasi
	Wheat	Rakh Dhiansar
	Pulse-based sequence	
Mung bean	Mustard	Agra
	Safflower	Akola, Bijapur
	Pigeon pea	Akola
Cowpea	Ragi	Bangalore
	Safflower	Jhansi
	Mustard	Jhansi
Cluster bean	Safflower	Jhansi
	Mustard	Jhansi
Horsegram	Ragi	Bangalore
Blackgram	Mustard	Agra, Varanasi
Soyabean	Gram	Indore, Dehra Dun
	Barley	Dehra Dun
	Safflower	Indore
	Oilseed-based sequence	
Peanut	Safflower	Indore
Sesame	Gram	Varanasi
	Fodder-based sequence	
Maize (fodder)	Gram	Varanasi, Udaipur
	Mustard	Udaipur
Maize + cowpea (fodder)	Tobacco	Anand
	Cluster beans	Anand
Cowpea (fodder)	Safflower	Jhansi
	Barley	Jhansi
	Gram	Jhansi, Varanasi

1st crop	*2nd crop*	*Centre(s)*
	Fodder-based sequence	
Sorghum (fodder)	Mustard	Varanasi
	Gram	Jhansi
	Barley	Jhansi

Tillage: Off-season tillage helps to keep the soil open for better intake of rainwater and better weed control, particularly in the light-textured soils. In the USA, chemical tillage is often undertaken under conditions of moisture stress, to minimise evapotranspiration losses.

Fertilisers: Nitrogen should be placed in split doses for *kharif* (June-October) crops while phosphorus should be placed basally. For *rabi* (October-March) crops, deep placement of nitrogen is required. Introduction of a legume in the pre-season, either in a rotation sequence or in inter-cropping, would have a net gain of 10 kg N ha^{-1}. Since fertiliser is a costly input, farmers need training in more efficient use. Fertiliser and other external inputs must be made available at cartable distances for adopting dryland farmers to use.

Soil and water management: Soil conservation should be dovetailed with runoff management by providing dug-out ponds as well as waterways. Gully plugging and nalla (drain) building combined with checkdams have been shown to improve the water regime in a given area. This should be considered.

Agro-forestry: Marginal lands should be placed under forestry and sylvi-pastoral systems. *Luacaena leucocephalla* and *Stylosanthus* species have been found encouraging for that purpose.

Improved implements: Proven, bullock-drawn tools needs to be commercially manufactured. The over-riding requirement is quality control. Some examples are:

Implement	*Place of origin*
Bullock drawn ridger seed	Hissar
Seed drill	Anantapur
Serrated blade harrow	Akola
Bullock drawn bakhar-cum-fertiliser seed drill	Jodhpur
Bullock drawn seed-cum-fertiliser drill	Varanasi
Bullock drawn deep furrow seeder	Ranchi
Bullock drawn seed-cum-fertiliser drill	Bangalore

Use of such implements makes a large difference in plant population and fertiliser use efficiency.

FLOODS

Flood prone areas again can be classified, depending on the periodicity and fury of floods. In chronically flood-prone areas, steps should be taken not only for minimising flood incidence through appropriate civil engineering works but also for making the

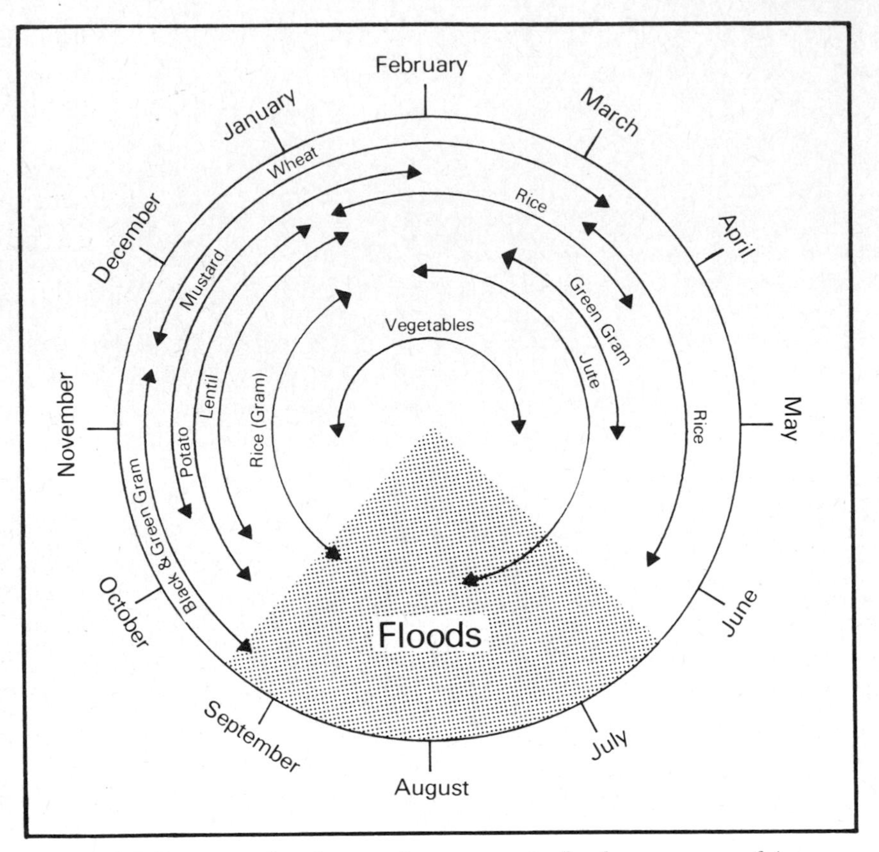

Figure 3.7. Restructuring the cropping patterns in flood-prone areas of Assam, India

flood-free season a major crop production season. The possibilities of undertaking a restructuring of crop patterns in the Brahmaputra Valley of Assam in India are shown in Figure 3.7.

BUILDING A NATIONAL FOOD SECURITY SYSTEM

A strong national food security system, integrating action to reduce the extent of crop production loss caused by unfavourable weather, can alone ultimately help to reduce the prospect of famine.

The steps involved in building an enduring national food security system have been described recently (Swaminathan, 1981). National food security means not only building grain reserves, but also paying concurrent and coordinated attention to:

Ecological security: Steps for achieving ecological security would include measures to protect the basic assets of agriculture and to minimise the liabilities. This can be achieved through the establishment of a National Land Use Board, which could foster, through appropriate scientific analysis and public policies, land and water use practices compatible with the concept of sustainable development. However, ecological security cannot be promoted by government alone. It has to be a joint sector activity involving the people and government agencies. Local level eco-development associations involving schools and colleges should be organised. Such associations could operate waste exchanges, to collect and recycle all organic waste. The economic benefits from eco-development and waste recycling could provide the motivation necessary for attracting public attention and participation. For example, in the United States it has been calculated that an estimated 300 million trees could be saved annually if the amount of paper recycled were trebled. Besides steps at the national level, regional and global level action plans to conserve our genetic and environmental heritage need to be developed. Some steps, such as the organisation of an International Bureau of Plant Genetic Resources, have already been initiated but many more are needed.

Technological security: Growth with stability should be the aim of agricultural development programmes. These would call for the breeding of high yield-cum-high stability crop varieties, with similar measures in animal husbandry and fisheries. The research system should be capable of promoting appropriate early warning-cum-timely action programmes. Pest survey and surveillance systems as well as soil, water, plant, and animal health care programmes will have to be developed. Here again, the total involvement of the local farming community will be necessary. Since, in the case of pest epidemics, political boundaries may not always provide the basis for appropriate early warning and control systems, it will be necessary for all the countries concerned with a problem to form regional control grids. The FAO-sponsored programme on locust control is a good example of the value of such cooperation.

Post-harvest technology and building grain reserves: A mismatch between production and post-harvest technologies frequently results in losses to both the producer and the consumer. Therefore, all operations after harvest, such as drying, storage, processing, transporting, and marketing will need to receive integrated attention. Depending upon possibilities, every country should maintain a reasonable grain reserve. For example, India derived immense benefit during the unprecedented drought year of 1979 from the grain reserve of about 20 million tonnes built by the government. A decentralised strategy of grain storage, under conditions of a free market economy, would also help to prevent panic purchases when conditions for crop growth are unfavourable and distress sales by poor farmers with no holding capacity when harvests are good. In fact, a decentralised plan for storing grain as well as water at appropriate locations all over the country, should be an essential element of the food security system of nations whose agricultural fortunes are closely linked to rainfall distribution.

Social security: In many developing countries, grain surplus and widespread hunger tend to co-exist. Even when there is food in the market, the lack of purchasing power leads to undernutrition and malnutrition. Therefore, in countries where agricultural production keeps ahead of population growth, the food and nutrition problem could be better stated in terms of mandays of employment rather than in metric tonnes of foodgrains. Right-to-work should become an integral part of a plan for food security. The integration of employment generation as an explicit aim in land and water use plans assumes relevance. Social security measures should include programmes such as "food for work" for able-bodied persons and "food for nutrition" for young children, pregnant and nursing mothers, and old and infirm persons. Rural development programmes should be designed to provide minimum needs in the fields of drinking water, education, health and environmental sanitation. In addition, there has to be a detailed manpower planning and employment generation strategy for rural areas based on a careful analysis of the possibilities for:

— land and water-based occupations, such as agriculture, horticulture, animal husbandry, fisheries and forestry;
— non-land occupations, such as small and village industries as well as agro-processing enterprises; and
— provision of relevant services.

In developed countries, the services sector tends to provide great opportunities for employment. In fact, as agriculture advances, more and more persons working on a daily wage status tend to be employed in the secondary and tertiary sectors of the economy. Historically, a rising standard of living has depended on the ability of agriculture to release manpower to other, more industrial pursuits.

Social security measures are needed as much for farmers and fishermen as for consumers. Suitable devices to insulate farmers against losses due to factors beyond their control will have to be developed. Old age pension and insurance measures to prevent the suffering of small farmers and agricultural labour when they become old and infirm are necessary.

Drinking water supply and nutrition education: Even when people have the requisite purchasing power, several forms of nutritional diseases arising from specific causes such as Vitamin A induced blindness, iron deficiency anaemia, goitre, etc., may still occur. Such nutritional disorders, attributable to well-identified causes, can be easily eliminated within a specific time frame through concerted efforts in nutrition education and intervention. Drinking water supply and mycotoxins in food need particular attention.

Population stabilisation: Depending upon the situation prevailing in each country with regard to the population-natural resources equation, an appropriate population policy will have to be developed and implemented. In countries like India, planning for economic development will be a futile exercise without the widespread adoption of the small family norm. Ecological security cannot be achieved, without arresting the rapid growth of both human and animal populations. Developments in preventive and curative medicine and improvements in environmental sanitation and safe drinking water supplies will lead to a continuous improvement in the average life span of humans. Family planning programmes will have to become an integral part of national food security systems. Livestock reform to regulate the pressure of cattle, sheep, goats, and other animals on the native vegetation is equally essential. In many developing countries, non-edible plants are becoming the dominant constituents of the local vegetation, since animals do not allow edible plants to survive. In some areas, non-edible shrubs are also cut for fuel. When this happens on a large scale, desertification results.

EMERGING SCENARIO

Specifications to the plant breeder

Detailed research on the interaction between climate and crop production is now gaining momentum. Considerable data are available on such crops as rice (IRRI, 1976 and 1980). We need similar studies on all major crops. More detailed characterisation of the environment in relation to the growing conditions for different crops is necessary. This will help in the formulation of more effective and appropriate research and development priorities and strategies. A rice weather project has recently been jointly sponsored by WMO and IRRI. Under this project, meteorological equipment will be installed at a number of selected sites of the IRTP irrigated nurseries. The locations selected will cover a wide range of climatic parameters, particularly with regard to solar radiation and minimum temperatures. The parameters to be measured on a daily basis are:

- solar radiation;
- maximum and minimum temperature;
- dry and wet bulb temperature;
- wind speed;
- rainfall.

Data from such studies will help provide detailed specifications to the plant breeder on the characteristics essential for better performance under conditions of moisture stress and moisture excess. An example of the power of modern plant breeding to develop varieties with wide adaptation and multiple resistance to pests is the history and characteristics of rice variety IR36, which now occupies over 10 million hectares in Asia and which has contributed to stabilising rice production to some extent.

Planning plant breeding programmes for stability of performance

(The history of IR36 was prepared by Dr. G. S. Khush of IRRI, under whose leadership the variety was developed).

History of Development of IR36 at IRRI

The initial crosses which led to the development of IR36 were made in June and July 1969. An early maturing line, IR579-48-1-2, from the cross of IR8/Tadukan, was crossed with another early maturing line, IR74B2-6-3, from the cross of TKM^2/TN1. These parents were selected because of their resistance to bacterial blight and stemborers and their excellent grain quality. In addition, IR747B2-6-3 is resistant to brown planthopper. The F_1 of this cross No. IR1561 was planted in July 1969 and the F_2 in December of the same year. F_3, F_4 and F_5 pedigree rows were grown in March, July, and November 1970, respectively. Many of the F_5 progenies were resistant to bacterial blight, stemborers, and brown planthopper but were susceptible to green leafhopper, tungro, and grassy stunt.

In May 1969, *Oryza nivara*, a wild species from India, was found to be resistant to grassy stunt. It was crossed to IR24 in July 1969. Since *O. nivara* is very poor agronomically, 3 backcrosses using IR24 as the recurrent parent were made, in October 1969, in June 1970, and in October 1970. Each generation of backcross progenies was screened for resistance to grassy stunt. Grassy stunt-resistant progenies from the third backcross (IR1737) looked essentially like IR24 and were flowering in the field in February 1971. F_5 progenies from IR1561 also were flowering in the field at the same time. A cross between IR1561-228-1-2 and a plant of IR1737 was made in February 1971. This cross was assigned IR2042. The F_1 seedlings of this cross were inoculated with grassy stunt and resistant seedlings were transplanted in the greenhouse in August 1971. An F_1 plant of this cross was topcrossed with gall midge resistant line, CR94-13, from CRRI, Cuttack, India. This line obtained in June 1971 was also found to be resistant to green leafhopper and tungro.

The F_1 seeds of this topcross were sown in January 1972 and seedlings were again inoculated with grassy stunt. Resistant seedlings were transplanted in the greenhouse and seeds from these plants were harvested in April 1972.

The F_2 population of this cross was grown without insecticide protection in July 1972 at Maligaya Rice Research and Training Center (MRRTC), Muñoz, Nueva Ecija, in Central Luzon, Philippines, because tungro disease pressure was very high in Central Luzon during 1971 and 1972. Moreover, stemborer populations at Maligaya in general are higher than at IRRI. The F_2 population was rogued at maximum tillering stage to remove tungro susceptible plants. Plants showing severe stemborer damage also were rogued. At maturity, 937 plant selections were made. These plant selections were grown at IRRI as F_3 progeny rows in the pedigree nursery planted in December 1972 without insecticide protection. There was severe build up of brown planthopper in the nursery and susceptible rows were killed.

All the F_3 rows were also inoculated with bacterial blight in the field at the maximum tillering stage. Susceptible rows were discarded and susceptible plants from the segregating rows were rogued. F_3 rows were also evaluated for blast resistance in the blast nursery and for resistance to green leafhopper and brown planthopper in the greenhouse. At maturity (March 1973), plant selections were made from the pedigree rows having multiple resistance to blast, bacterial blight, green leafhopper, and brown planthopper.

A special field screening technique was devised to screen F_4 progenies for resistance to grassy stunt. A field was planted to IR24, which is highly susceptible to brown planthopper and grassy stunt. No insecticides were used and the population of brown planthopper built up rapidly. Since there was a lot of inoculum of grassy stunt on IRRI farm, all the plants of IR24 were infected with grassy stunt. One-meter-wide strips of plants in this field were cut and a seedbed was made in the strips. F_4 progenies of IR2071 were sown in these seedbeds in May 1973. The young seedlings were inoculated with grassy stunt by the viruliferous insects abundant on the older seedlings of IR24 which moved to young seedlings in the seedbed. The seedlings were transplanted in the field in June 1973. Progeny rows which were susceptible to grassy stunt showed 100% susceptible plants. These F_4 rows also were inoculated with bacterial blight in the field and evaluated for resistance to blast in the blast nursery and for resistance to green leafhopper and brown planthopper in the greenhouse. An early maturing line numbered IR2071-625-1 appeared to be very vigorous and it was resistant to grassy stunt. Data in the field book showed that it was also resistant to blast, bacterial blight, green leafhopper, and brown planthopper. Seeds of this line were bulk harvested at maturity in early September along with many other lines of this cross.

A small seed increase plot of IR2071-625-1 was planted in September 1973. This also was entered in the replicated yield trials in January 1974. The seed increase plot planted in September 1973 matured in February 1974 and 400 individual plant selections were harvested. These plant selections were planted to headrow blocks at the end of February 1974. Plot No. 252 looked uniform at maturity and was bulk harvested in May 1974. It was labeled IR2071-625-1-252. This reselection from the original line was entered in the replicated yield trial in June 1974 and was tested in other coordinated trials.

By the end of 1973, tungro incidence in the Philippines declined and the breeding materials could not be screened for tungro resistance under field conditions. Therefore, seeds of F_5 lines of IR2071 cross, including IR297-625-1 and several thousand other breeding lines, were sent to

Indonesia in January 1974 for planting at Lanrang in South Sulawesi where tungro incidence was very high. IR2071-625-1 was found to be resistant to tungro in this trial.

Seeds of promising F_6 lines were planted without insecticide protection at CRRI, Cuttack, in July 1974 for screening for resistance to gall midge. Gall midge pressure at Cuttack during months of September and October is generally very high and these materials were exposed to heavy gall midge pressure. IR2071-625-1-252 was found to be resistant to gall midge. Meanwhile, F_6 lines were screened for resistance to stemborer in the screenhouse at IRRI during the July-October 1974 growing season. IR2071-625-1-252 was found to have best resistance to stemborer of all the lines of this cross.

By the end of 1974, multiple resistance of IR2071-625-1-252 to blast, bacterial blight, grassy stunt, tungro, green leafhopper, brown planthopper, stemborer, and gall midge was established. Two seasons' yield data in replicated yield trials during 1974 indicated that this line had very high yield potential. The analysis of grain quality showed that it had excellent, long, slender and translucent grains with high milling recovery. Therefore, IR2071-625-1-252 was entered in the Philippine Seedboard Lowland Cooperative Performance Trials in the 1st season of 1975. This line outyielded all the other entries in the early maturing group of these trials during two seasons of 1975. At its March 1976 meeting, the Rice Varietal Improvement Group of Philippine Seedboard recommended the naming of IR2071-625-1-252 as IR36. This recommendation was approved by the Philippine Seedboard at its 21st annual meeting in May 1976 and IR36 became a seedboard variety. It replaced IR26, the dominant variety at that time, within a year. More than 50 per cent of the total rice area of the Philippines has been planted to IR36 since 1978.

IR36 was first entered in the International Rice Yield Nurseries (IRYN-E) in 1975. It topped the list of 16 entries tested at 24 locations in nine countries. Similarly it was the highest yielding entry in the 1976 IRYN-E consisting of 20 entries tested at 37 locations in 16 countries. The test entries came from several national rice improvement programmes. IR36 has been used as an international check in the IRYN-E since 1977 and has consistently been among the three top yielding entries.

IR36 was entered in the All-India Coordinated Rice Improvement Project (AICRIP) trials in 1976 and was the top yielding entry among 16 entries tested at 56 locations throughout India in 1978. It was released for cultivation by the Central Variety Release committee of India in 1981.

To summarise, 13 varieties from 6 countries are involved in the ancestry of IR36. These parents contributed various traits for its genesis. Wide adaptation and high yield potential were contributed by Peta and IR8, respectively. Early maturity and superior grain quality came from TKM6 and Tadukan. Resistance to bacterial blight and stemborers also was inherited from TKM 6 and Tadukan. *Oryza nivara* contributed genes for resistance to grassy stunt and blast. Resistance to tungro, green leafhopper, and gall midge was inherited from Ptb 18 and Ptb 21. IR36 was the first improved variety of rice to have multiple resistance to all the major diseases and insects (Table 3.8).

Because of its wide genetic background, genes for tolerance to various problem soils, such as salinity, alkalinity, iron toxicity, zinc deficiency, iron deficiency, and aluminium toxicity (Table 3.9) were combined from different parents. It may have inherited drought tolerance from *Oryza nivara.*

The estimated area planted to IR36 in countries of Asia is given in Table 3.10.

That history of one recent variety of rice shows how a scientific programme for minimising fluctuations in yield caused by soil factors and pest incidence can be developed. Progress in biotechnology research should further enhance the power of scientific plant breeding because of the opportunities provided by tissue culture for screening and for alien gene transfer.

Table 3.9. Reaction of some IR varieties to adverse soil conditions (scale 0-9)[1]

	Wetland soils							Dryland soils	
	Toxicities					*Deficiencies*		*Toxicities*	*Deficiency*
	Salt	Alkali	Peat	Iron	Boron	Phosphorus	Zinc	Aluminium and manganese	Iron
IR5	4	7	0	6	4	5	5	5	4
IR8	3	6	5	7	4	4	4	4	4
IR20	5	7	4	2	4	1	3	4	4
IR28	7	5	6	4	4	3	5	5	6
IR36	3	3	3	3	3	7	2	4	2
IR42	3	4	5	3	2	3	4	5	6
IR48	4	7	5	6	0	5	5	3	4

1. 0: no information; 1: almost normal plant; 9: almost dead or dead plant.

Table 3.10. Estimated area planted to IR36 during 1981

Country(s)	Area (million hectares)
Indonesia	5.3
Philippines	2.3
Vietnam	2.1
India, Malaysia, Laos, Kampuchea, Sri Lanka, Bangladesh	>1.0
Total	>10.7

ISSUES RELATING TO ATMOSPHERIC CARBON DIOXIDE PROBLEM

There is also a need to look carefully at certain long term trends, for example the increase in carbon dioxide (CO_2) concentration in the atmosphere (Kellogg and Schware, 1981). By the middle of the next century, the continued burning of fossil fuels as a source of energy is likely to result in a doubling of the CO_2 content in the atmosphere relative to the amount present in 1860. The present CO_2 level of about 335 parts per million per volume (ppm) is expected to increase to about 380 ppm by the end of the century (Bach, 1981). Such an increase will have two types of consequences — on photosynthesis, because of the greater quantity of carbon available to plants from the atmosphere, and on climate. Computer models indicate that, on a global basis, average temperatures on the earth's surface will rise 2 to 3°C with a doubling of atmospheric CO_2. Both evaporation and precipitation may increase by about 9 per cent.

Given adequate solar radiation, soil nutrient availability, and irrigation, increased atmospheric CO_2 should act as a fertiliser for crop plants, raising both photosynthetic production and water-use efficiency. Greenhouse experiments have indicated that a doubling of CO_2 under good crop management can increase biomass yields by about 40 per cent. Structural adaptations in farming systems will be necessary, both to take advantage of the favourable consequences of CO_2 increase and to face its negative repercussions. The CO_2 effects should be especially important for such crop plants as rice, wheat, millet, and potato which have a C_3 photosynthetic pathway. Corn, sugarcane, and sorghum, with a C_4 pathway, are likely to be limited by solar radiation and nutrient and moisture availability rather than by CO_2.

Plant breeders in the tropics should aim at developing varieties that will have higher net photosynthetic production and use less water as atmospheric CO_2 content increases. The strains should not respond to a warmer atmospheric temperature by an increase in respiration that would cancel out the effect of CO_2 fertilisation. There is an opportunity for accelerating attempts to increase total photomass production in the major crop plants.

Herman Flohn (1981) recently estimated the changes in average surface temperature and precipitation that may occur in different latitude belts if atmospheric CO_2 goes up to 560-580 ppm (i.e., about twice the nineteenth century value) (Table 3.11).

If these estimates prove correct, major changes in surface and underground water supply due to altered precipitation and evapotranspiration patterns in several parts of the world could occur. Some of the agriculturally productive areas of the USA, Canada and USSR may be adversely affected. The USA, USSR and China have about 90 per cent of

Table 3.11. Probable effects of increased CO_2 content (Flohn, 1981)

Latitude	Average annual change in surface temperature (°C)	Change in precipitation (percentage)
60°N	+7.5	+18
50°N	6	+4
40°N	+6	-14
30°N	+4.5	0
20°N	+2.5	+20
10°N	+1.5	+20
Equator	+3	0
10°S	+4	-20
20°S	+4.5	-5
30°S	+4	+5
40°S	+4	+12
50°S	+3	+12
60°S	+2.5	+12

the world's coal reserves. Since higher CO_2 concentrations may affect these countries adversely, they may be unwilling to develop an export trade in coal. This in turn will have implications for the energy-short countries (Revelle, 1981).

The projections made by Herman Flohn would imply more rain in some of the drought prone areas and more floods along the Ganga and Brahmaputra rivers in the Indian sub-continent. Expansion in major and medium irrigation works as well as extensive denudation of vegetation may also influence weather, particularly the micro climate, in different ways. Hence, the plant breeder, with the help of climatologists and environmentalists, will have to assemble diverse genotypes which will profit from increased CO_2 and precipitation or, alternatively, withstand the adverse impact of higher temperature and enhanced evapotranspiration.

Anticipatory breeding in this area will include steps for:

— developing strains which can help to enhance productivity per day and per litre of water;
— breeding of varieties which can help to tap the production potential of flood-free seasons in flood prone areas;
— taking advantage of opportunities for external trade that may emerge in case the traditional bread basket areas of the world find it difficult to sustain production at high levels;
— developing genotypes which can derive advantage from enhanced CO_2 availability; and
— conserving genetic variability and maintenance of genetic diversity in crop populations.

FORECASTS OF PEST EPIDEMICS

Satellite photography of cloud movements and remote sensing techniques are aiding in the development of early warning systems against pest incidence. Cooperation between meteorologists and plant and animal health care specialists will be very beneficial in refining such techniques.

CLIMATE AND INLAND AND MARINE FISHERIES

Extensive data are now becoming available on the implications of weather patterns to aquatic productivity. This area of research is particularly important to countries which have a large, exclusive economic zone.

PREDICTION OF CLIMATE CHANGE

Huke and Sardido (1980) have pointed out that there has been a significant increase in weather variability in India since the early 1950s. It appears that the benign climate of the late 1930s and 1940s has started to change. Whether this change will be for the better or for the worse is not yet clear. Studies of this kind, with a deeper analysis of likely causes for climate change, are particularly important for countries with expanding populations and shrinking land resources for agriculture.

CONCLUSION

Among all the factors that influence agricultural output, climate is by far the most significant. This is particularly because, to a considerable extent, climate-induced production variations are beyond the control of man. However, early warning-cum-timely action programmes can be initiated based on predictions of likely weather patterns. Contingency plans and compensatory production programmes to suit different weather probabilities can become integral parts of crop production planning. By learning to live in harmony with nature, the benefits of favourable climatic conditions can be maximised and the risks associated with aberrant weather minimised.

Data from space exploration have shown that we have only our own planet to depend upon for the food we need. If we safeguard the basic agricultural assets of soil, water, flora, fauna, and the atmosphere, the chance of achieving freedom from hunger for present and future generations will correspondingly improve. To achieve the requisite blend of professional skill, political will, and people's action for promoting development without destruction, it will be necessary for every country to have effective tools for programme formulation and implementation.

Developing countries possessing a large, untapped agricultural production potential but faced with problems of chronic food shortage should consider the establishment of a suitable machinery for preparing and implementing these action codes —

- *good weather code:* this code should specify measures for maximising production during normal and favourable climatic conditions. This code also should specify the action to be taken in arid, semi-arid, and drought prone areas and the steps that should be taken during a good season for strengthening the ecological infrastructure of such areas. Only once in several years will such areas receive good rainfall. That year should be used fully for extensive tree plantation, sand dune stabilisation measures, and, where appropriate, aerial seeding, etc.;
- *unfavourable weather code:* this should have the components to deal with problems of drought, floods, typhoons, etc. on the lines earlier discussed.

Such measures will help countries to be prepared for different weather patterns and to initiate anticipatory action to mitigate the effects of aberrant weather rather than to be content with palliative action after misfortune has set in. Then even calamities can be converted into opportunities for further development.

REFERENCES

Arnon, I. (1981) Modernization of Agriculture in Developing Countries. John Wiley and Sons, New York, p. 565.

Bach, W. (1981) "The CO_2 issue — What are the Realistic Options?" *Climatic Change* Vol. 3, pp. 3-5.

Biswas, A. K. (1980) "Crop-Climate Models: A Review of the State of the Art," in *Climatic Constraints and Human Activities* edited by J. Ausubel and A. K. Biswas, IIASA Proceedings Series, Pergamon Press, pp. 75-92.

Campbell, W. H. and Bryson, R. A. (1982) "Year-in-Advance Forecasting of the Indian Monsoon Rainfall," *Environmental Conservation,* (April, 1982).

Flohn H. (1981) *Major Climatic Events as Expected During a Prolonged CO_2-induced Warming,* Report prepared for Institute for Energy Analysis, Tenn., Oak Ridge: Oak Ridge Associated Universities.

Huke, R. E. and Sardido, S. (1980) "Climate Change in India," *Proceedings of Symposium on Agrometeorology of the Rice Crop,* International Rice Research Institute, pp. 173-180.

India, Government of/Directorate of Economics and Statistics, Ministry of Agriculture (1979) *1979 Indian Drought: The Challenge and Strategy to Combat it.* Mimeo, 37 pps and appendices. Ministry of Agriculture, Krishi Bhavan, New Delhi.

International Rice Research Institute (1980) *Proceedings of a Symposium on the Agrometeorology of the Rice Crop organised jointly by WMO and IRRI,* p. 256.

International Rice Research Institute (1976) *Proceedings of a Symposium on Climate and Rice,* p. 505.

Kampen, J., and Burford, J. R. (1980) "Production Systems, Soil-Related Constraints, and Potentials in the Semi-Arid Tropics, with Special Reference to India," In *Priorities for Alleviating Soil-Related Constraints to Food Production in the Tropics.* International Rice Research Institute, Los Baños, Philippines, pp. 141-165.

Kanwar, J. S., Virmani, S. M. and Kampen, J. (1981) "Management of Vertisols: ICRISAT Experience." *Paper prepared for presentation at the 1982 International Soil Science Congress held at New Delhi,* in February 1982.

Kellogg, W. W. and Schware, R. (1981) *Climate Change and Society: Consequences of Increasing Atmospheric Carbon Dioxide.* Westview Press, Colorato, USA, p. 178.

Mooley, D. A., Parthasarathy, B., Sontakke, N. A. and A. A. Munot, (1981) "Annual Rain-Water over India, its Variability and Impact on the Economy." *Jour. Climatology* Vol. 1, pp. 167-186.

Murthy, R. S. (1981) "Distribution and Properties of Vertisols and Associated Soils. Improving the Management of India's Deep Black Soils." Mimeo, 106 pps. The International Crops Research Institute for the Semi-Arid Tropics (ICRISAT), Patancheru, Andhra Pradesh 502 324, India.

National Defense University and US Department of Agriculture (1980) *Crop Yields and Climate Change to the Year 2000,* Vol. 1, 1980, p. 128.

Oldeman, L. R. and Frère, M. (1982) *A Study of the Agroclimatology of the Humid Tropics of Southeast Asia.* FAO/UNESCO/WMO Interagency Project on Agroclimatology, FAO, Rome, p. 229.

Pantastico, Ed. B. and Cardenas, A. C. (1980) "Climatic Constraints to Rice Production in the Philippines," *Proceedings Symposium on Agrometeorology of the Rice Crop, IRRI,* 1980, pp. 3-8.

Revelle R., (1981) "The Earth's Potential Land, Water and Energy Resources" *Nobel Symposium on Population Growth and World Economic Development,* Norway, September 7-11, 1981.

Rosenberg, N. J. (1980) Research in Great Plains Drought Management Strategies. University of Nebraska, USA.

Roy, P. K .(1981) *Agricultural Insurance.* Pergamon Press, London, p. 419.

Ryan, J. G. (1974) "Socio-economic Aspects of Agricultural Development in the Semi-Arid Tropics," Occasional Paper 6, Economics Programme, The International Crops Research Institute for the Semi-Arid Tropics (ICRISAT), Patancheru, Andhra Pradesh 502 324, India.

Stansel, J. W. (1980) "The Impact of World Weather Change on Rice Production," *Proceedings Symposium on Agrometeorology of the Rice Crop,* IRRI, 1980, pp. 143-151.

Schultz, T. (1953) *The Economic Organisation of Agriculture,* McGraw Hill, N.Y., p. 197.

Swaminathan, M. S. (1979) "Evolution of Drought Management in India." Paper presented at the International Symposium on hydrological aspects of droughts. New Delhi, Dec. 3-8, 1979. Mimeo, 18 pps. and appendices. ICAR, Krishi Bhavan, New Delhi.

Swaminathan, M. S. (1981) "Building a National Food Security System." *Indian Environmental Society,* p. 138.

Swaminathan, M. S. (1982) *Science and Integrated Rural Development.* Concept Publishing Company, p. 354.

CHAPTER FOUR

Climate and Water Resources

Asit K. Biswas
International Society for Ecological Modelling

CLIMATE AFFECTS WATER resources most directly and immediately in terms of precipitation. Temperature and wind also effect water availability through the process of evapotranspiration. The principle of mass conservation can be successfully used to determine the amount of water available for use by the terrestrial ecosystem. This means that the rate of change of storage of water in water bodies, soil and groundwater in an area, over a specific time interval, is equal to the rate of water supplied from the sky in the form of precipitation minus the combined outflow from that area in terms of runoff and evapotranspiration. Continuous replenishment of water in surface-water bodies, both natural and man-made, and as groundwater is essential for long-term human survival.

It is not possible to design a sustainable development process within a region which can only use the annual replenishment of water for that specific year. This is due to the great variation in annual rainfall from one year to another. Thus, during a period of low rainfall, more water may have to be used than is being replenished, by using water stored during above-average rainfall years. If water consumption is continually higher than the surface and groundwater storage, it will eventually contribute to supply exhaustion.

History has many examples where water consumption patterns were higher than supplies available. When this happens, either development has to be abandoned in that area or consumption-supply ratio has to be brought into balance as quickly as possible, often at great hardship to the people of that area. Probably the best example of such a case is what happened to Akbar the Great, the most famous of the Mogul emperors of India, when he decided to establish a new capital for his vast empire. The best architects available were asked to design a magnificent palace at Fatehpur Sikri, not far from Agra, in the dry plains of Northern India. The cream of artisans worked for several years to complete the capital, and a vast amount of resources was spent on the realisation of the emperor's dream.

As any modern traveller to Fatehpur Sikri will attest, it is an excellent testimonial to Indian architecture. The completed palaces are still intact, untouched by the intervening centuries.

The history of the new capital, however, was not so auspicious. Akbar used it only for 15 years, and then he was forced to abandon it rather ignominiously to return to the old capital. The main reason for this unsustainable development was due to the fact that the rate of water consumption in such an arid climate was far higher than the rate of replenishment. Thus, when the available water resources of the area were exhausted, the emperor had no alternative but to abandon his expensively-built capital.

Thus, one of the main issues facing water resources planners is how to design water developments that are sustainable over the long-term. The reliability of precipitation, to some extent, depends on the climate. The problem is especially difficult in arid, semi-arid and monsoonal climates, which can have extraordinarily wide variations in precipitation from one year to another. The problem is further aggravated by the fact that areas of low precipitation are also areas where the variation is great. In contrast, year-to-year variation in areas having moderate marine climates could be fairly small. Figure 4.1 shows the global distribution of percentage variability of annual precipitation (Landsberg, 1975).

CLIMATE AND WATER DEVELOPMENT

From a climate-water development viewpoint, there are four important issues which should be noted.

First, all water development projects are designed on the basis that climate of the recent past is also the key to the future, at least so far as the design-lives of projects are concerned. Generally, on the basis of 10 to 25 years of available data on hydrological variables such as precipitation, streamflow, etc., attempts are made to estimate probable extreme events in terms of low and high flows (droughts and floods). Water resources systems are then designed so as to minimise the impacts of the extreme events within an acceptable economic cost. It is thus implicitly assumed that the climatic and hydrologic regimes of the past will continue in the future.

The assumption of the constancy of the climatic regime, at least on a long-term basis, is incorrect. Climate has never been stable or constant throughout the earth's history, and there is absolutely no reason to believe that the situation will change. Thus, strictly speaking, the question of whether climatic change is taking place is tautological, since climate has continued to change as long as the earth has had an atmosphere, and will probably continue to do so even after man has disappeared from the earth, and thus no longer able to record those changes. Much uncertainty exists at present about the future climate since scientists cannot agree at present on causes, effects and trends.

Furthermore, human activities are becoming increasingly important, especially in terms of having perceptible impacts on climatic variables. For example, it has been conclusively proven that human activities can alter climatic norms on a meso-micro scale, but considerable uncertainty exists in making scientifically accurate estimates of potential climatic consequences due to man-made causes. Landsberg (1979) has provided an excellent view of the effects of human activities on climate.

Figure 4.1. Global distribution of per cent variability of annual rainfall (Landsberg, 1975)

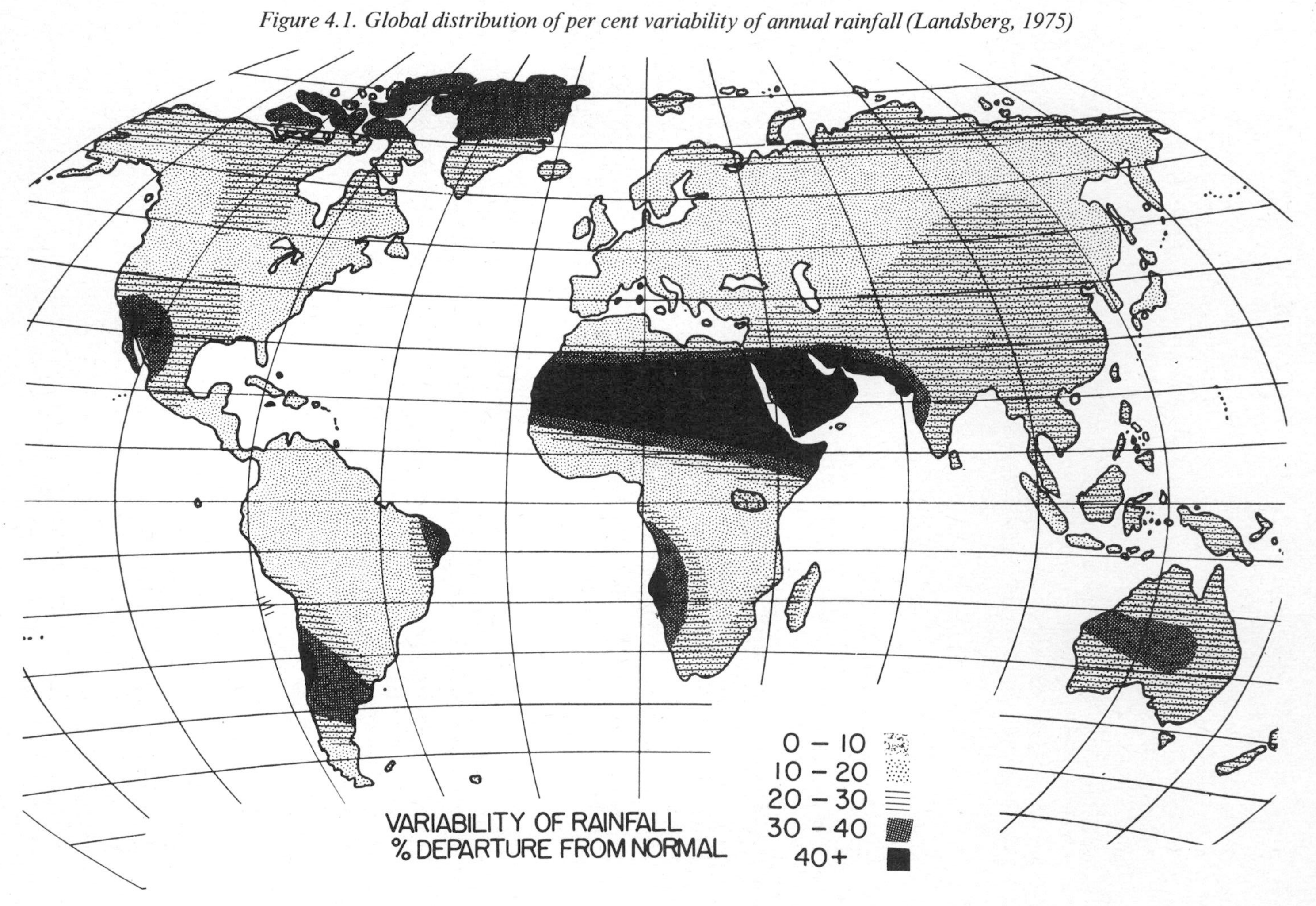

It should also be noted that human activities can alter the hydrological regime as well. For example, large-scale deforestation or afforestation in the upper reaches of rivers may alter the patterns of runoff. Again it is not possible to make reliable estimates of what changes may occur in the flow regime due to changes in vegetative covers upstream.

Thus, even though the question of climatic change is tautological on a long-term basis, what needs to be considered within a water resources planning framework is the probability of such a change occurring within the design life of water resources systems. Normally, water resources systems are designed for an economic life, spanning over 50 to 100 years, though some large ones could be for as long as 500 years. In addition, often due to large investment necessary, and complexities arising out of the preparation of socially, economically, technologically and environmentally acceptable plans, an additional period of 10 to 30 years is often necessary to prepare adequate plans, have them accepted by appropriate decision-making processes, and then have the plan implemented. Thus, from a pragmatic, water resources development viewpoint, it would be desirable to have climatic forecasts, ranging from 60 to 130 years, depending on a variety of factors like the type and location of hydraulic structures, proximity to the centres of population, standard of safety necessary, development purposes etc. It is important, however, that for design and operational purposes, the climatic forecasts should be more site-specific, a point often forgotten in recent literature. Forecasts of global climatic "averages" or trends, while interesting, are of little value in the actual planning and design of water resources systems.

There are basically two schools of thought concerning the incorporation of climatic variability into water planning. The first school suggests that since the economic life of most water resources structures is between 50 to 100 years, and since there has been no evidence of climatic change during the past 200 years, the probability of a major change during the next 200 years is minimal, and therefore the whole question is somewhat academic. Thus, Chin and Yevjevich (1974) suggested that "since most systems have been built with the economic project life in the range of 40 to 100 years, the chances are minimal that the expected natural water supply would be significantly different during these life spans than in the past 200 years .... This question is, however, not crucial for the next several generations of contemporary earth population, but rather is more of an academic interest like many other human concerns with the long-term future." They further attempted to show that climatic fluctuations could be reduced to a deterministic component based on the Milankovich theory of astronomical cycles and a simple Markovian stochastic component.

Others have taken a contrary view, and have pointed out that climatic variability is an established fact. Accordingly, the variability should be recognised, and should be analysed and used in the water resources planning and management process. Mitchell *et al.* (1975) clearly state that "The lessons of history seem to be that climatic variability is to be recognised, and dealt with as a fundamental quality of climate, and that it should be potentially perilous for man to assume that the climate of future decades and centuries will be free from similar variability."

The importance of considering climatic variability has been clearly demonstrated, especially in terms of water resources management, by O'Connell and Wallis (1973).

They showed that the reservoir firm yield, assuming a 50-year design life, could have different estimates even when Markov and other persistent generating mechanisms used yielded samples having identical expected values for the mean, variance and lag-one correlation. In other words, the analysis clearly indicated that it is not only important for hydrologists and water planners to understand the nature of climatic variability and persistence but also imperative that such considerations be incorporated in the planning process. In another paper, Wallis and O'Connell (1973) showed that statistical analyses of hydrologic data of average period of years would usually lead one to believe that a Markov generating mechanism adequately represents the streamflow pattern in a real world. Such analyses could instill a false sense of security amongst water resources planners, since they are likely to be erroneous. This is because various statistical tests carried out do not have the power to distinguish between samples of such lengths taken from Markovian and more persistent generating mechanisms.

In addition to the above consideration, there could be another factor which could enhance the economic uncertainty, especially in terms of available future water supply, much of which will be necessary to increase agricultural production through irrigation and better water control. As the National Research Council (1977) of the United States has noted, the uncertainty is due to the "threat of climatic change, real or imagined, which can alter what farmers will plant, as well as what economists think they will plant. At present, we do not know whether climatic change will result in more or less water available behind a proposed dam. Further, we do not know what the net effect of a postulated average annual global temperature change may be on the future water requirements within an actual specific project area. It would appear that rather than global averages, the climatic-change estimates that might actually influence projections of water use are estimtes of changes in frequency and severity of extreme periods. Climatologists may find reliable, nonstationary, extreme value forecasting beyond their ability."

The second important issue is that while there may be major fluctuations in precipitation in an area from year to year, changes in runoff generally tend to be more pronounced than precipitation in dry years. Since most evapotranspiration requirements have to be met prior to surface runoff and groundwater recharge, and evapotranspiration tends to be relatively constant from one year to another, any downward drift in precipitation often tends to result in even more reduction in surface runoff. Thus, in the areas that depend exclusively or primarily on water supply from rivers, the problem of water shortage could be more serious than the low precipitation figures may indicate.

Changnon (1977) has provided an example of this issue by analysing 55 years of streamflow data for the Illinois River basin. He considered the lowest flow periods for five distinctly separate 30-month periods during 1915-70. He assumed evapotranspiration to be constant by using an average figure. The analysis indicated that the lowest 30-month rainfall, 173 cm, was associated with the lowest streamflow, 13 cm, but whereas the precipitation figure was a 24 percent departure from normal, the corresponding departure for streamflow was 79 percent (Table 4.1).

The third important issue is the fact that water development projects are generally designed to operate under a wide range of climatic fluctuations. It is a common

Table 4.1. Comparison of precipitation and runoff during 30-month dry periods in Central Illinois River Basin, 1915-70 (Changnon, 1977)

Rank	Precipitation (cm)	Runoff (cm)	*Percent of normal*		*Percentage departures below normal*	
			Precipitation	Runoff	Precipitation	Runoff
1	173	13	76	21	24	79
2	183	15	80	27	20	83
3	201	28	89	46	11	54
4	208	33	91	55	9	45
5	213	36	93	57	7	43

practice amongst water planners to estimate maximum probable flood so that the hydraulic structures designed can withstand them, and also to determine lowest flows to be expected over a specific period of time, generally 3 to 5 years, during which required water can be supplied from the system. The final decision as to the ranges of the extreme climatic regime within which the system can function effectively is dependent on economics, institutional and legal requirements and societal concern. Thus, it can be argued that because of this built-in flexibility, water development projects can better withstand not only climatic fluctuations but also some modest changes in climatic and hydrological regimes, either natural or man-made, than many other areas of human activities like agricultural production, which have been discussed in other chapters, where such changes or fluctuations could have catastrophic effects.

It should, however, be remembered that while much scientific progress has been made during the last two decades in estimating extreme climatic events, both in terms of magnitude and probability of occurrence, we still cannot predict them with any significant degree of confidence (Biswas, 1971). Thus, considerable uncertainty still exists in the design of water resources systems to withstand extreme events.

Equally, it should be noted that continuing population growth, increasing urbanisation, and expansion of agricultural development into marginal areas in recent years, have generally tended to increase the potential socio-economic impacts of variations of water availability compared to similar impacts in the past. For example, comparatively small changes in climatic variables, especially in terms of precipitation and temperature, could mean the difference between abundant agricultural production or failure in marginal areas. Thus, in many cases, the flexibility available to decision-makers or range of errors permissible are diminishing, which means the potential for social and economic vulnerability to specific segments of society is increasing. Such considerations are especially important in the arid and semi-arid areas of developing countries. In the future, with further increases in population, urbanisation and use of marginal land, the socio-economic impacts of variability of water sources are likely to be even more serious and crucial than at present (Biswas, 1979).

The fourth important consideration is the fact that there are important differences due to climatic factors in different parts of the world so far as water development activities are concerned. Some of these differences are serious and others may be subtle, but together they may require different design, planning and operational practices for water resources systems from one area to another, often within the same country. The environment and institutional processes within which water development takes places could also be different. Consequently, water management process can differ from one climatological area to another.

Some of the major implications of climate-development interrelationships have already been discussed in Chapter 1, many of which are also relevant for water resources development. In this section some of the different aspects of lake and reservoir management in the tropics and temperate regions will be considered as an example of variable management requirements in such areas.

Tropical and temperate zone rivers may differ widely in such factors as flow rate, turbidity, pollutants present in water and temperature. Naturally the quantity and quality of water flowing into lakes or reservoirs determine many of their characteristics. For example, sedimentation is a major concern in most tropical reservoirs. Heavy tropical rainfall, the onset of rain after the summer when vegetative cover is generally at its least, and the soil is very dry and loose, and steep valley slopes all contribute to high sedimentation rates. Poor management of upland watersheds due to deforestation and overgrazing further aggravates the situation. As the silt-laden river water enters the reservoir, the velocity of flow is reduced and thus the rate of sedimentation increases. Accordingly, special attention needs to be given for sedimentation management.[1]

Sedimentation reduces storage capacity, and hence the useful life of a reservoir. It also results in the loss of nutrients, which otherwise may have reached agricultural and aquatic ecosystems downstream. Clear water coming out from the reservoir often increases erosion of the river banks and deltas. High rates of sedimentation also can have adverse effect on the spawning of fish and distribution of biota. Presence of suspended materials increases turbidity, thus significantly reducing photosynthesis, which in turn reduces productivity of higher trophic levels (NRC, 1982).

If trees and vegetation within a reservoir are not cleared prior to inundation, water quality implications could differ between the tropics and temperate climates. Decay of organic material is much faster in the tropics than in temperate climates, and rates of oxygen consumption during biochemical degradation and nutrient release have different impact magnitudes on water quality in the two regions. In cold temperatures, submerged trees can last a long time. For example, in the Gounin Reservoir in Canada, tree trunks show very little deterioration after being flooded for some 55 years (Biswas, 1982).

Infestation of aquatic weeds is often a serious problem in the tropics and semi-tropics. Major problems with aquatic weeds have been observed in places as diverse as Aswan, Kariba, Nam Pong (Thailand) and Brokopondo (Surinam). Once infestation starts, growth of aquatic weeds can be very fast. For example, *Eichornia crassipes,* commonly known as water hyacinth, covered an area of about 50 square kilometres in Lake Brokopondo within the short period of February to December, 1964. In little

1. For a comprehensive review of reservoir sedimentation problems from different parts of the world, see Biswas, 1982.

over two years, by April 1966, it had covered more than 50 per cent of the surface of the reservoir, an area of about 410 square kilometres. Similarly, in Egypt, weeds were not a problem until 1964. However, in 1965, water hyacinth started to spread prolifically in Middle Egypt and the Nile Delta area. By the beginning of the spring of 1975, various types of aquatic weeds, sometimes mixed with dense algae, had invaded more than 80 per cent of all the watercourses and a great part of the Nile itself. Experience in the Congo Basin has been somewhat similar. Between 1952, when the weeds were first observed, and 1955, they spread a distance of some 1,600 kilometres, covering large areas of the Congo River. The problem of aquatic weeds is further aggravated by the presence of irrigation and drainage channels.

Weeds create several problems. First, water losses are greatly increased by their evapotranspiration process. If irrigation is practised, more water has to be released to ensure adequate water availability in the lower reaches. These two factors often account for a tremendous amount of water loss. Thus, at Aswan it has been estimated that 2,875,000,000 cubic metres of water are lost per year due to the preceding two factors alone. A better perspective can be obtained if it is considered that at present prices, 2,875,000,000 cubic metres of annual over-year storage on the upper reaches of the Nile at Aswan will cost some 216,000,000 US dollars, which is a not insignificant figure.

Weeds also tend to increase the incidence of diseases such as malaria and schistosomiasis by providing a favourable habitat for invertebrate vectors and intermediate hosts (mosquitoes and aquatic snails) for disease-causing agents. They create further problems by interfering with the operation and maintenance of hydroelectric generation and pumping stations, and by competing with fish for space and nutrients.

Control of aquatic weeds in the tropics and semi-tropics, especially after invasion, is a difficult and expensive task. Mechanised or manual clearing of weeds, especially in shallow waters, has been quite successful, but in deep waters it is not a very viable alternative. Weeds thus removed can be used to produce animal feed, biogas or manure. In certain countries, e.g. China, aquatic plants are specially cultivated as animal feed.

Chemical herbicides have been used extensively to control weeds, but chemical control is not very effective for submerged weeds. In addition, herbicides often pose a major environmental hazard to aquatic organisms and deteriorate water quality; their long-term effects on aquatic ecosystems and human health are little understood.

Another type of control is biological: fish, snails or aquatic grasshoppers are introduced to control weeds. There is still much to be learned about the use of biological controls. Naturally, the three control measures are not mutually exclusive; they are often used in various combinations for optimal weed control. The type of control measures to be used depends on various local situations such as the type of weed, density of infestation, depth and width of the channel, time of application, water-use pattern, proximity of crop areas sensitive to herbicides, availability of material from local or foreign sources and availability of skilled manpower.

Water-development schemes have often enhanced or created favourable ecological environments for parasitic and water-borne diseases such as schistosomiasis, dengue and dengue haemorrhagic fever, liver-fluke infections, bancroftian filariasis, and malaria. These diseases are not new; for example, schistosomiasis was known during

Pharaonic times. However, the incidence of these diseases is much higher if extensive perennial irrigation is practised. Schistosomiasis is currently endemic in over 70 countries, and affects over 200 million people. The same number are at present infected with malaria in the tropics and subtropics, and another 250 million are infected with bancroftian filariasis. Similarly, plant growth around reservoirs provides a suitable habitat for tsetse fly to transmit trypanosomiasis to humans and domestic animals (M.R. Biswas, 1979).

In contrast to the diseases mentioned, water resources developments tend to reduce the incidence of onchocerciasis. The intermediate host, the *Similium* fly, tends to breed in fast-flowing waters, which are often drowned by the construction of dams. Thus, the construction of the Volta dam destroyed the breeding ground of the *Simulium* fly that existed upstream. However, adequate measures should be taken to ensure that new breeding places do not develop, especially in the fast-flowing water near spillways.

Water quality considerations are also somewhat different in tropical reservoirs when compared to their temperate counterparts. Higher average daily temperature and thermal discharges from thermal or nuclear power plants, if present, can have adverse water quality impacts. This is because of two reasons. First, dissolved oxygen levels in water decrease with increases in temperature. For example, oxygen solubility from air into water is 16 per cent lower at 30°C compared to 20°C, and is 29 per cent lower at 40°C than 20°C. Second, the rates of biological processes increase with temperature, and accordingly biochemical oxygen demand (BOD) increases as well. The net result of these two effects could be a significant reduction in the dissolved oxygen level in water, which may have deleterious effects on fisheries and other components of an aquatic ecosytem (Biswas, 1980). Thus, both reservoir and fisheries management may need to be more finely-tuned in the tropics than in temperate climates.

Climate fluctuations and water resources planning and management

Considering the significant potential impacts of climate on water resources planning and management, it is really surprising that not much work is being done in this important area. Recent activities in the area of water resources planning and management will be discussed under the following three headings: study of cycles, synthetic hydrology and other techniques.

Study of cycles: If the present methods and techniques for water resources planning and management are critically reviewed, it soon becomes evident that one of the most important basic philosophies, albeit somewhat implicit, is that the recent past is the key to the near future. It is relatively simple to consider the near future, the length of which is the design life of the proposed system. However, problems start to arise when a review of the past is attempted — since the techniques for doing so have not been universally agreed to, and thus have resulted in some controversy.

The situation becomes even more complicated when one tries to predict the near future on the basis of limited data on the past, a common occurrence in most of the developing countries and in many instances in the developed world. Often hydrologists consider themselves to be lucky to have reliable data for only 10 to 15 years. Considering the inherent problem of sampling errors in such short records, it is indeed difficult to make predictions for the future. The problem becomes even more complex if it is assumed that the climatic and/or hydrological regimes have changed to a new state in recent years. In this case, it would mean that the observed data of the past belong to one statistical population, and the events of the near future will belong to another. This means that one of the fundamental assumptions of hydrology becomes immediately invalid, and the recent past could no longer be considered to be the key to the future.

Facing these types of difficult problems, some scientists have considered the possibility of using cycle analysis as a basis for climatic forecasts. The technique is a little more sophisticated than trend analysis. Basically, cycle analysis is an empirical method wherein reasonably long periods of observed data of different phenomena are analysed to determine the existence of cycles. These cycles are then used as a form of forecasting algorithm to predict the future. Cycle analysis has always been popular among certain scientists and they have often used it as a forecasting tool. In fact, this form of analysis has been popular enough to support its own society, devoted to the study of cycles, and sponsor its own speciality journals, e.g. *Interdisciplinary Cycle Research.* In all cycle analysis, non-cyclic independence of data is implicitly assumed. This, as Mandlebrot and Wallis (1969) have pointed out, is somewhat unlikely for climatic data.

In the area of climatic analysis, considerable emphasis has been placed in recent years to the study of solar cycles, and their apparent correlation to water availability. These studies range from the precipitation characteristics of Addis Ababa and Asmara, to variations in the water levels of Lake Victoria, to droughts in the high plains of the United States. In each of these cases, attempts were made or are being made to correlate certain climatic events with sunspot cycles, with inconclusive results.

Attempts to correlate water-related phenomena with sunspot cycles have been made for decades. However, without scientific explanation of interrelationships between sunspots and climatic events, it is rather easy to get into trouble. For example, as early as 1937, Stetson wrote a book on *Sunspots and their Effects,* in which he shows some "interesting" correlation between sunspot numbers with Dow Jones Stock market averages, rabbit population, building contracts or number of automobiles. These are shown in Figure 4.2. Without a physical understanding of the interrelationships between the events being correlated, statistical analyses could often easily result in spurious correlation.

One could argue that even between seemingly unrelated events such as sunspot numbers and rabbit population, one could find some physical linkages. For example, the output of the sun obviously has some effect on the climate, though precisely what it is, is difficult to say. If, as some researchers have claimed, there is some relation between sunspot numbers and droughts, sunspots could have some impact on vegetation, and through vegetation on the rabbit population. Such linkages, however, are hypothetical, and have yet to be proved with any degree of confidence. In the absence of any realistic scientific theory, such speculations are interesting conjunctures at best, and nonsense at worst.

Thompson (1973) has analysed double sunspot cycles with droughts in the state of

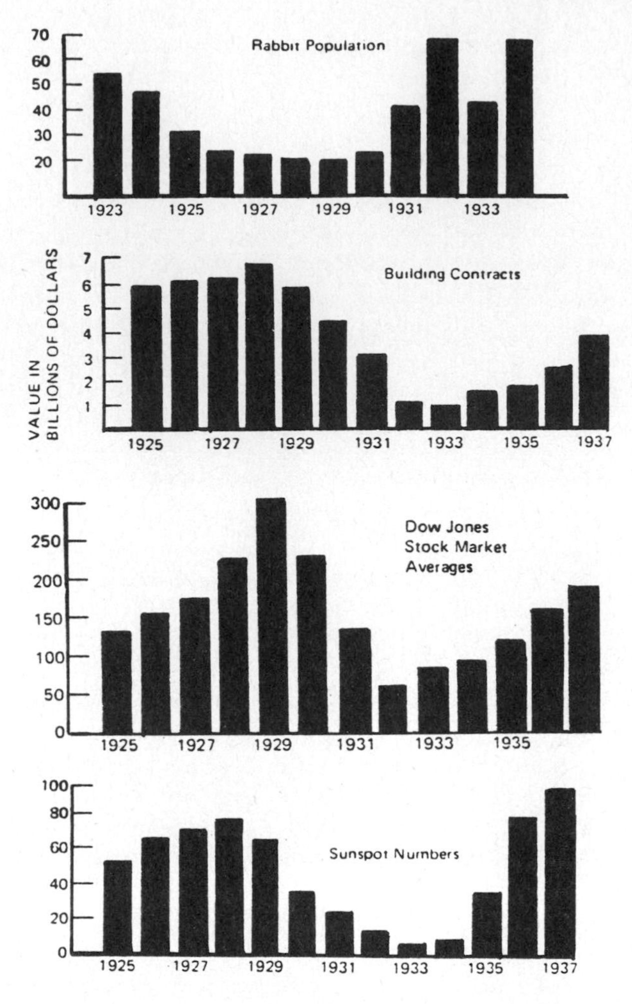

Figure 4.2. Correlation between sunspots, rabbit population, building and the Dow Jones index (from Stetson).

Nebraska in the United States. It appears that droughts have occurred in Nebraska during alternate minima of the double sunspot cycle. If the result is to be taken at its face value, it would seem that the droughts occur in 22-year cycles, centred near the minima of that circle.

There are many questions to be answered before the 22-year cycle of droughts in Nebraska can be taken seriously. For example, why do such cycles occur in the plains of the United States, but not elsewhere? Why do the droughts not affect the entire plain? It seems that they occur in different parts of the plain, and the cycles do not necessarily affect the similar parts of the plain each time. Nor does their timing of occurrence coincide exactly with alternate sunspot minima. Without reasonable explanations of such anomalies, the 22-year cycle is likely to remain an interesting speculation.

With regard to cycle analysis, it is interesting to note that in 1960, an attempt was made to forecast precipitation, for the period 1961 to 1970, for 32 locations within the United States. This was done by developing an algorithm based on solar cycles of 91, 46 and 23 years, and 91, 68, 55, $44\frac{1}{4}$, $38\frac{1}{2}$, 34, $30\frac{1}{3}$, $25\frac{1}{2}$, 21, 11.87, 11.29, 9.79 and 8.12 months (Abbot, 1960; Abbot and Hill, 1967). A similar algorithm was used to predict temperature in ten locations for the 1962 to 1967 period (Abbot, 1961) of the United States.

The advance ten-year forecasts for precipitation were then compared with actual figures. Using Spearman's rank correlation coefficient to determine the accuracy of the advance forecasts, it was found that nearly half of the correlations were positive and the balance was negative. This means that forecasts were no better than random. An analysis of the forecast temperatures was not much better.

In the absence of better scientific explanation and documentation, hydrologists should realise that forecasts based on cycle analysis, like many other climatic forecasts, are often based on "poor foundations, apparent similarities, parallel-looking curves and analogous trends" (Gani, 1975). Further, it can be experimentally verified that "stochastic processes with low-frequency components can yield sample functions with numerous 'significant' cycles are often induced into the analysis of moving average manipulations to smooth what are otherwise noisy data" (National Academy of Sciences, 1977; Slutzky, 1927).

Use of paleoenvironmental indicators: Some progress has been made in recent years in estimating long-term climatic fluctuations on both global and hemispheric scales. Since such estimates are of little direct value to water resources planning, it is necessary to interpret and translate past climatic variability in terms of regional hydrologic variables like streamflow.

Not much work in this area has been carried out thus far. Among the few scientists working in this area, Stockton (1975, 1977) and Stockton and Jacoby (1976) have successfully reconstructed total annual runoff data, by using tree-ring information for the Upper Colorado Basin. Figure 4.3 shows reconstructed hydrographs for total annual runoff for two rivers: Green River — a tributary river within the Upper Colorado River Basin, at Green River, Utah; and the Colorado River at Lee Ferry, Arizona. From the reconstructed hydrographs, several important points emerge. For example, if a hydrologist were to analyse only the streamflow data of the two rivers, based on actual observations, then it is highly likely that rather high estimates of mean annual flow and variance would be obtained. This is because during the early part of the present century, both the rivers experienced high flows — over a prolonged period — which, as the reconstructed records indicate were the longest for the 450 years of reconstructed history. In other words, without such reconstruction of streamflow data, one would have obtained much higher water availability from both rivers, than would have been actually the case. Furthermore, analysis of the long-term hydrograph of the Colorado River indicated that the series is significantly different from random and can be best modelled by a mixed autoregressive-moving average scheme

Figure 4.3. Reconstructed hydrographs for total annual runoff for Green River at Green River, Utah, and Colorado River at Lee Ferry (Stockton, 1975)

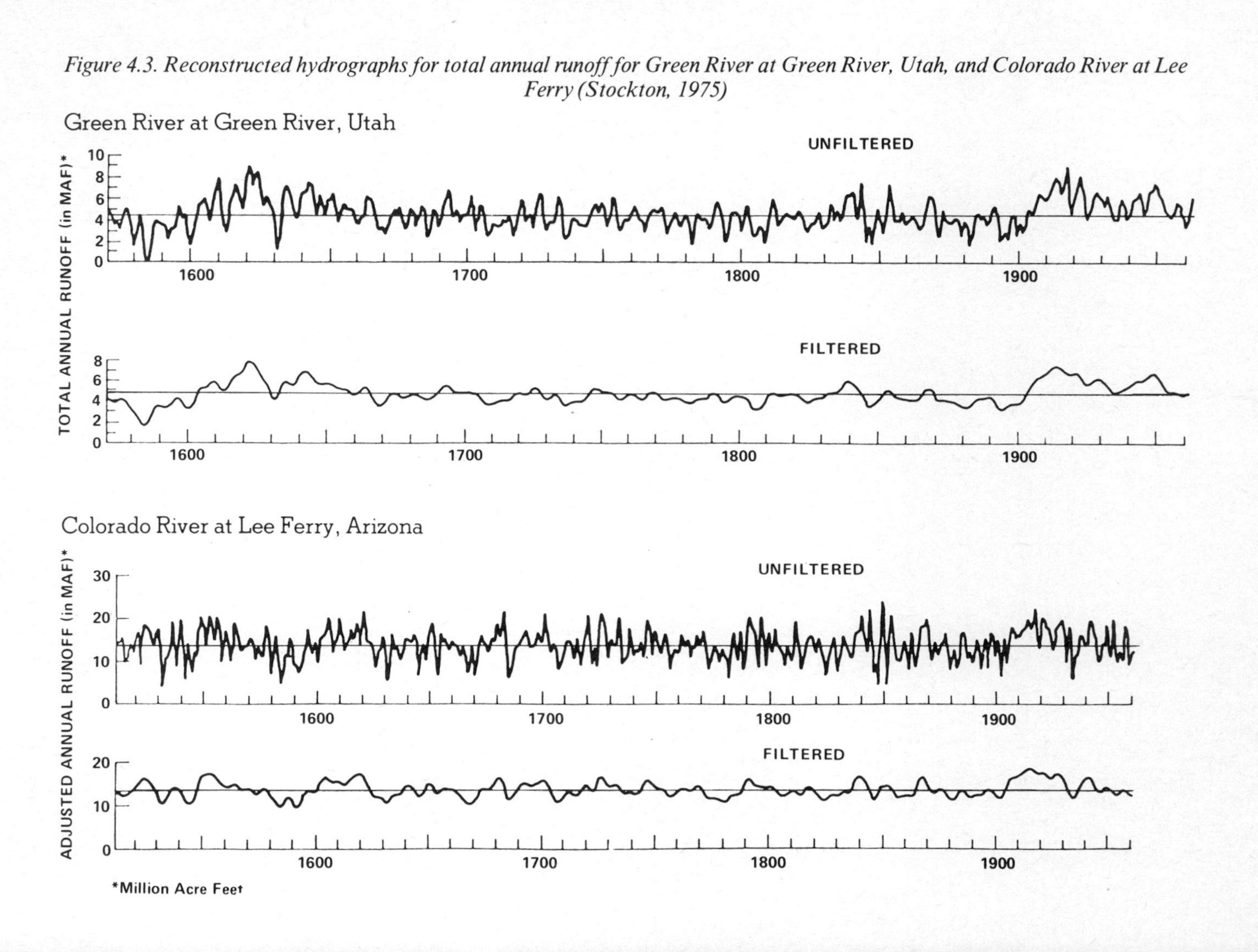

(Stockton, 1977). Also, the Colorado did not exhibit the long-term persistence found in the reconstructed history of the Green River.

Similar reconstructions of hydrological time series data are not readily available, but Landsberg *et al.* (1968) have reconstructed annual temperature and precipitation totals along the eastern seaboard of the United States, centred in Philadelphia, for the period 1738 to 1968. The preliminary reconstruction of this 230-year data shows an anomalous wet period for about 50 years during 1830 to 1880, and a trend in the annual temperature data caused primarily due to the "lack of cold years since the turn of the 20th century".

Use of paleoenvironmental indicators for actual water resources planning is still in its infancy, and much work remains to be done to develop new techniques and refine the existing ones. There is, however, no doubt that this is a promising area of work, and expenditure of further resources is likely to pay rich dividends.

Synthetic hydrology: There are a few studies that are being currently conducted in the area of synthetic hydrology, with regard to climatic fluctuations.

In some pioneering work, the Harvard University Water Group used in 1962 synthetic lag-one Markov generating mechanism of the type:

$$x_1 = p_1 x_{i-1} + E_i$$

where E_i are independent random variables scaled to preserve the observed means, variances and lag-one correlations, P_1. The problem with such a generating mechanism is that it tends to give lower critical flow values than those of the observed sequences rather than providing values that are equally likely to occur. In order to improve analytical process, there has been a tendency to introduce incremental changes like elimination of biases in the parameter-estimating procedures, or even use of multi-lag models. These changes, however, have not solved the basic problem.

Some attempts have been made to introduce certain aspects of climatic changes in the lag-one Markov model. For example, Schwarz (1977) assumed that variations of monthly streamflow can be represented by a skewed logarithmic lag-one Markov model, and that four parameters, mean, standard deviation, and coefficients of skewness and serial correlation, can define such a model. These parameters were then changed according to a long-term climatic change model, which gave some indication of the system response due to effects of climatic changes.

Schwarz used this concept to see how the streamflow records of the Potomac River at the Points of Rock gauge may be affected. The drainage area at that locale is 24,996 square kilometres. He used 75 years of streamflow data (1897-1970) to estimate the logarithmic mean, standard deviation and coefficients of skewness and serial correlation for each month. From these parameters, a 1,000 year synthetic flow record was generated. One set of statistics was then varied to generate eight alternative "climatically-changed" 1,000 year records.

From this analysis, Schwarz concluded that climatic change does not seem to "radically alter the water-supply planning process". Based on this analysis, Schwarz has also provided a speculative impact matrix of climatic change as shown in Table 4.2.

Schwarz's attempt to generate synthetic streamflow, as a surrogate for climatic change, is a valiant effort in a view direction, and it could possibly give planners a

Table 4.2. Speculative Impact Matrix of Climatic Change (Schwarz, 1977)

	Parameters of Climatic Change				
Attributes of water supply systems	A Decrease in mean streamflow	B Increase in variance of streamflow	C Increase in skew of streamflow	D Increase in persistence of streamflow	E Speed with which change occurs
1. Yield from unregulated streams	Some effects, but likely not very large except if change in mean is large or combined with other changes	Severe effects; however, generally short term	Significant effects because number of days of low flow increase relative to few very high flow periods	Significant effects more through duration of low flows than severity	Not applicable
2. Yield from reservoirs	Significant to severe effects particularly if reservoirs develop a high percentage of the average flow	Medium to no effects depending on the size of the reservoir in relation to drainage area; larger reservoirs will suffer smaller effects	Medium to no effects depending on the size of the reservoir in relation to drainage area; larger reservoirs will suffer smaller effects	Significant to severe effects especially if res-reservoir long-term storage is limited	Not applicable
3. Yield from groundwater	Significant in the long run, especially if draft on aquifer is near average recharge	Little if any significance	Little if any significance	Effects severe and of long duration	Not applicable
4. System reliability	Some effects, other than effects accounted for under 1-3	Some reduction due to constant change in flows in addition to effects under 1-3	Little or none, other than effects under 1-3	Little or none, other than effects under 1-3	Sudden changes severely affect reliability, slow ones less or not at all
5. Magnitude and control of demand	No significant effect	No significant effect; often recurring short-term restrictions may reduce their effectiveness	No significant effect	No significant effect; emergency restrictions likely to become less effective over long droughts	Significant and visible effects, relatively fast changes could force major steps toward conservation and demand control
6. Cost of operation of water system	No significant effects except for additional construction that might eventually ensue to alleviate long-term shortages	Possible increase due to turbidity, increased pumping between systems if applicable; possible additional reservoir construction	No significant effects likely	No significant effects except search for new sources	No effects
7. Pressure on and ability of the water system to respond to change	Pressure for expansion would be created if shortages occur repeatedly; ability to respond would not be affected by hydrologic event	Pressure for expansion would be created, but rapid return to normal may for some time inhibit expansion	Pressure for expansion would be created if shortages occur repeatedly; ability to respond would not be affected by hydrologic event	Pressure for expansion would mount over time and increase likelihood of action; however, long high flow periods may inhibit development	Sudden or relatively near future changes could increase action; long-term changes (20 years+) even if known would likely be ignored by existing institutions

"feeling" of the system sensitivity to change. Such analyses have many major disadvantages, some of which have been pointed out by Schwarz himself. First, the relation between climatic variations and streamflow at a specific site is still largely undefined. Second, no estimate of likely climatic change, even on a global basis, is available at present. The situation could be even more difficult to construct a site-specific master long-term climatic change model, based on the physical and meteorological levels, which could then dictate changes that are to be analysed. Thus, in the study mentioned, the investigators were forced to make arbitrary variations in the parameters, without having much information on the probability of occurrence of such changes. Third, it is difficult to select a good generating algorithm. Currently, several models exist, but each have their own advantages and disadvantages. Schwarz's approach was a nonlinear, non-stationary, synthetic hydrology, which is theoretically interesting but not very useful for water resources planning purposes.

Other techniques Not many new concepts have been put forward to incorporate the effects of possible climatic change in water resource planning and management in recent years. In fact, after an extensive literature search and a questionnaire survey of leading research institutions, only one technique was found that could be termed relatively "new". This is the introduction of the concept of robustness and resilience within the context of mathematical modelling of water resources systems.

Literally hundreds of models have been developed in recent years in most parts of the world to transfer climatic functions such as precipitation and temperature to streamflow. Use of such climatic transfer functions, which implicitly assume that climate will continue to be as before in the real world, has not been all that successful. For example, the World Meteorological Organisation carried out an evaluation of the performance of ten models on up to six watersheds for a period of two years. The results were often greatly in error, even when recorded daily precipitation and temperature values were used for the basins considered. Individual events sometimes showed little or no correlation between observed and forecast values, and the sum of the forecast flows for the 2-year period ranged in excess of 40 per cent. This, of course, is not surprising, especially when the inherent uncertainties associated with hydrological techniques currently being used for parameter estimation are considered. Thus, there is a hierarchy of uncertainties, in depending on which parameters are being estimated. For example, there is some uncertainty associated with the determination of streamflow, more uncertainty with the estimation of the standard deviation and even more uncertainty with the higher moments of the low probability density functions.

If this is the state of the art for climatic transfer function models, all of which are constructed with the explicit assumption that model parameters are time invariant, i.e. streamflow is a stationary process, the situation is far worse, when the development of a realistic model with time variant parameters is considered. Matalas and Fiering (1977) have addressed to some of these problems.

Matalas and Fiering have attempted to introduce the concept of robustness and resilience in hydrological analyses. None of these two concepts is new: they came from statistics and ecology respectively. Robustness is defined as "the insensitivity of the system design to errors, random or otherwise, in the estimates of those parameters affecting design choice." It was suggested that some designs have built-in buffering, and consequently are more robust than others in that they are applicable over a wider range of population mean values. In contrast, some designs could be optimal over a narrow range of the population mean. Normally, as a rule, it can be said that large systems have substantial robustness built-in, in that they are technologically and institutionally capable of adopting to larger stresses. Matalas and Fiering suggest:

> Part of the design problem is to identify the types of climatic shift that might be anticipated and to determine if they are sufficiently precipitous with respect to flow characteristics to dictate a change in system design. It is not necessary for this purpose to know or to try to determine whether there is a true climatic shift. This may be an interesting scientific question, important in its own right, but it is virtually meaningless for the design of water-resource systems. It is also unimportant to know if the population moments of the flow distribution are modified, because, again, while this might be an important hydrologic matter, it is important for water-resource design if, and only if, the changes, when coupled with economic criteria, lead to new design.

If a water resources system has been designed to perform at an optimal level, and has adequate built-in buffering, it should be capable of being operated at another system level. Such a deviation of the system from an optimal level will undoubtedly mean some economic loss, but this can be constrained for design purposes at a certain percentage — which can then be referred to as the resilience at that level. Such a design will not be optimal — in the standard sense of the term — but could possibly be the operational choice, which can be expected to perform reasonably well under changing climatic conditions. In other words, a system designed for a specific climate can be modified to operate under a different set of climatic norms, at a certain economic cost. This flexibility then becomes the resilience of the system. It should be noted that resilience so defined cannot only be enhanced by changes in structural measures, but also by using other means like zoning, insurance, subsidies and price structures.

The technique proposed by Matalas and Fiering seems to have considerable future potential. However, from a strictly water resources planning viewpoint it is unlikely to be used at present. Much work needs to be done before the technique can be considered for actual use in the planning of water systems.

In another development, WMO and UNEP cooperated in evaluating climate and water resources for agricultural development, for the Sudano-Sahelian zone of Africa (WMO, 1976). These studies are important in the sense tht they provide comprehensive analysis of the available data on different climatic variables — precipitation, temperature, wind speed and direction, vapour pressure, solar radiation and evapotranspiration. They point out some of the problems of analyses of climatic data. For example, thirty years of records at a single station may not be a good parameter for rainfall for planning considerations, especially in the northern half of the Sudano-Sahelian zone. Such averages of annual rainfall, even when considered over a long period of years, may provide insufficient and unreliable information. Throughout this zone, monthly averages or averages over even shorter periods, often show considerable inconsistencies that can probably be explained by the randomness of the local rainfall. For these reasons, it was suggested that median was a more useful parameter of rainfall rather than normal.

CLIMATE AND WATER QUALITY

Concern about dispersal of pollutants through climatic factors has continued to increase since the late 1960s. Air pollutants are dispersed far and wide through the atmosphere. While such dispersal reduces the concentration of pollutants as they are transported from the point of origin, they may nevertheless precipitate in the form of acid rain, often as far as hundreds or even thousands of kilometres away from the sources, depending on the strengths of prevailing wind. Thus, the water quality of precipitation has increasingly become an important concern during the last decade, especially in developing countries.

Many pollutants have been observed to have been precipitated. They range from metals such as zinc and lead to various chemicals such as pesticides. The concern with acid rain, however, has centred on two man-made compounds — sulphur oxides (SO_x) and oxides of nitrogen (NO_x). These compounds react in the atmosphere to form sulphuric acid and nitric acid, which then precipitate on the earth's surface in both rain and snow.

Emissions of sulphur oxides can be controlled, but such controls are expensive. Control requirements differ radically from one country to another. Technology that can be applied for removing oxides of nitrogen from industrial stack gases does not exist at present. A direct result of such emissions has been lower pH values of rainfall, which indicates higher acidity. For example, the pH values of rainfall over large areas of southern Sweden and Norway have been observed to have fallen from a normal value of 5.7 to around 4.5-4.2.

Acid rain can have adverse effects on aquatic ecosystems, terrestrial biota and soil. A survey of over 1,500 lakes in south-western Norway indicated that over 70 per cent of the lakes having a pH value lower than 4.3 contained no fish (Barney, 1980). Similar results have been noted in Sweden, Canada and the United States. Forest growths in some areas have also suffered from acid rain. Soil acidification may also result from acid rain, depending on the amount of calcium present in the soil. Growing acidification of soil can only reduce agricultural yield. Lime needs to be applied to the soil to neutralise its acidity.

OTHER CONSIDERATIONS

Extreme climatic events such as floods and droughts have serious social and economic impacts on countries. For example, according to WMO (1975), 22 countries in the Asia and Pacific region sustained a damage of 9.9 billion US dollars only from typhoons and floods, an amount which was almost as large as the World Bank loans to these countries during the same period.

For efficient operation and management of water resources systems, it is essential to have more information on the magnitude and probability of occurrence of extreme climatic events. This means it is essential to monitor climatic variables on appropriate time and space scales to develop a better data base for both design and management purposes, and to develop better theoretical basis for such designs. It is equally necessary to ensure that the information collected is available not only to water resources' planners and managers, but also to other people who need to use it, e.g. farmers. If it can be predicted with some degree of reliability that a growing season is likely to be dry, farmers can substitute some crops that can better withstand dry seasons than others. Under such circumstances, sorghum can be a good substitute for corn. However, if the prediction turns out to be wrong, and the season turns out to be average, farmers will incur economic losses for having planted sorghum instead of corn. In other words, it does not pay to play guessing games with climate: the results could be either excellent or catastrophic.

REFERENCES

Abbot, C.G. (1961) "A Long-Range Temperature Forecast", *Smithsonian Miscellaneous Collections,* No. 143, p. 5.

Abbot, C.G. (1960) "A Long-Range Forecast of United States Precipitation", *Smithsonian Miscellaneous Collections,* No. 139, p. 9.

Abbot, C.G., and Hill, L. (1967) "Supplement to a Long-Range Forecast of United States Precipitation", *Smithsonian Miscellaneous Collections,* No. 152, p. 5.

Barney, Gerald O. (Study Director), (1980) *The Global 2000 Report to the US President: Entering the 21st Century,* Vol. 1, Pergamon Press, Oxford, p. 185.

Biswas, Asit K. (1982) "Environment and Sustainable Water Development", Key-Note Lecture for the Fourth World Congress of International Water Resources Association, in *Water for Human Consumption,* Tycooly International Publishing Ltd., Dublin, pp. 375-392.

Biswas, Asit K. (1980) "Non-radiological Environmental Implications of Nuclear Energy", *Environmental Conservation,* Vol. 7, No. 3, pp. 229-237.

Biswas, Asit K. (1979) "Management of Traditional Resource Systems in Marginal Areas", *Environmental Conservation,* Vol. 6, No. 4, pp. 257-264.

Biswas, Asit K. (1971) "Some Thoughts on Spillway Design Flood", *Hydrological Sciences Bulletin,* Vol. 7, No. 6, pp. 63-72.

Biswas, Margaret R. (1979) "Environment and Food Production", in *Food, Climate, and Man,* Edited by Margaret R. Biswas and Asit K. Biswas, John Wiley & Sons, New York, pp. 125-158.

Changnon, S.A. (1977) "Climatic Change and Potential Impacts on Water Resources", in *Proceedings of the Symposium on Living with Climatic Change: Phase II, Reston, Virginia, November 9-11, 1976,* Report MTR-7443, Mitre Corporation, McLean, Virginia, pp. 85-93.

Chin, W.Q., and Yevjevich, V. (1974) "Almost — Periodic, Stockstic Process of Long-Term Climatic Changes", *Hydrology Paper* No. 65, Colorado State University, Fort Collins.

Gani, J. (1975) "The Use of Statistics in Climatological Research", *Search,* Vol. 6, No. 11-12.

Landsberg, H.M. (1979) "Effects of Man's Activities on Climate", in *Food, Climate and Man,* Editors: Margaret R. Biswas and Asit K. Biswas, John Wiley & Sons, New York, pp. 187-236.

Landsberg, H.E., (September 23, 1975) "Weather, Climate and Settlements", Background Paper for United Nations Conference on Human Settlements, Vancouver, Canada, 31 May-11 June 1976, Report A/CONF. 70/8/1, United Nations, New York, 65 pp.

Landsberg, H.E., Yu, C.S., and Huang, L., (1968) "Preliminary Reconstruction of a Long Time Series of Climatic Data for the Eastern United States", *Applied Mathematics Technical Note* BN-571, Institute of Fluid Dynamics, University of Maryland, 30 pp.

Mandlebrot, B.B., and Wallis, J. R. (1969) "Some Long-Run Properties of Geophysical Records", *Water Resources Research,* Vol. 5.

Matalas, N.C., and Fiering, M.B. (1977) "Water-Resource System Planning", in *Climate, Climatic Change and Water Supply,* Studies in Geophysics, National Academy of Sciences, Washington, D.C., pp. 99-110.

Mitchell, S.M., *et al.* (1975) "Variability of the Climate of the Natural Troposphere", *Climatic Impact Assessment Program, Monograph 4,* Department of Transportation, Washington, D.C.

National Research Council (NRC), Committee on Selected Biological Problems in the Humid Tropics, 1972, *Ecological Aspects of Development in the Humid Tropics,* National Academy Press, Washington, D.C., pp. 176-226.

National Research Council (1977) *Climatic Change and Water Supply,* Studies in Geophysics, National Academy of Sciences, Washington, D.C., 132 pp.

O'Connell, P.E., and Wallis, J.R. (1973) *Choice of Generating Mechanism in Synthetic Hydrology with Inadequate Data,* International Association of Hydrological Sciences, Madrid Symposium.

Schwarz, H.E. (1977) "Climatic Change and Water Supply: How Sensitive is the Northeast?" in *Climate, Climatic Change and Water Supply,* Studies in Geophysics, National Academy of Sciences, Washington, D.C., pp. 111-120.

Slutzky, E. (1927) "The Summation of Random Causes as the Source of Cyclic Processes", *Econometrika,* Vol. 5, p. 105.

Stetson, H.T. (1937) *Sunspots and their Effects,* McGraw-Hill Book Co., New York.

Stockton, C.W. (1977) "Interpretation of Past Climatic Variability from Paleoenvironmental Indicators", in *Climate, Climatic Change and Water Supply,* Studies in Geophysics, National Research Council, National Academy of Sciences, Washington, D.C., pp. 34-45.

Stockton, C.W., 1975, "Long-Term Streamflow Construction in the Upper Colorado River Basin Using Tree Rings", in *Colorado River Basin Modelling Studies,* Edited by C.G. Clyde, D.H. Falxenborg, and J.P. Riley, Utah Water Research Laboratory, Utah State University, Logan, Utah, pp. 401-441.

Stockton, C.W., and Jacoby, G.C. (1976) "Long-Term Surface Water Supply and Streamflow Levels in the Upper Colorado River Basin", *Lake Powell Research Project Bulletin.*

Thompson, L.M. (May 9, 1975) "Weather Variability, Climatic Change and Grain Production", *Science,* Vol. 188, No. 4188, pp. 435-541.

Wallis, J.R., and O'Connell, P.E., "Firm Reservoir Yield: How Reliable are Historic Hydrological Records", *Hydrological Sciences Bulletin,* Vol. 18, p. 347.

WMO (1976) "An Evaluation of Climate and Water Resources for Development of Agriculture in the Sudano-Sahelian Zone of West Africa", *Special Environmental Report* No. 9, WMO, Geneva, 289 pp.

WMO (1975) "The Role of Meteorological Services in the Economic Development of Asia and South-west Pacific", *Report 422,* WMO, Geneva, p. 69.

CHAPTER FIVE

The Effect of Climate Fluctuations on Human Populations: a Case Study of Mesopotamian Society[1]

Douglas L. Johnson and Harvey Gould
Graduate School of Geography, Clark University, Massachusetts

IN NEARLY ALL human-environment systems there is a tendency for population growth to press against the limits established by technologically exploitable resources. Few systems are so stable in their population dynamics that variation in population level does not occur. Historically, there is evidence for slow, progressive growth in total global population, a trend that has accelerated dramatically in the last three centuries. Yet in contrast to this cumulative expansion of global population is the evidence for periodic collapse of populations in particular areas. The remains of these defunct civilisations litter the archaeological landscape, and arouse speculation as to the causes that produced such catastrophic results.

One recurrent conjecture for such collapses has been exogenous environmental causes. In the case of Mycenean civilisation, Bryson and Murray (1977) invoke shifts in the average position of winter storm tracks, which makes less moisture available to the cities of the Peloponnesus, as an explanation for the decline and ultimate collapse of Mycenae. McNeill (1977) indicts the transport of disease from one ecological setting to another as an important population destabilising variable whenever previously isolated disease pools are brought into contact. The unfamiliar infectious disease that crippled Athens in 430-429 BC during the Peloponnesian War, the possible outbreak of measles and smallpox in the Antonine plague of AD 165-180 and its successor in AD 251-266, and the weakening of Byzantine and Persian empires as a consequence of repeated outbreaks of bubonic plague after AD 542 before the Muslim Arab invasions are intriguing examples of the interaction between disease, population dynamics and history. A similar argument for population decline in the Islamic portions of the Middle East during the fourteenth century due to the Black Death is made by Dols

1. This paper is based on research partly supported by the National Science Foundation, Grant No. ATM 77-15019. The invaluable contributions of Robert W. Kates, Richard A. Warrick, Richard Hosier and Robert Obeiter in the preparation of this paper are also gratefully acknowledged. Nonetheless, the conclusions and opinions expressed herein remain the responsibility of the authors.

(1977). Another example of exogenous causation is the decline of urban civilisation in the Indus Valley after 1800 BC. This decline may have been due to destructive flooding caused by geomorphological changes (Wheeler, 1968) as well as the removal of vegetation in the foothills of the Himalayas.

External causation is only one part of the process of environmental and social change over long time periods. Internal causes also play a major role, and it is important to emphasise the interactive and multi-causal nature of environmental change. In this chapter, our intention is to explore a case study situation in which not only internal and external causal factors can be differentiated, but also in which changes can be observed over a long time sequence of human occupation. In many areas it is possible that important relationships between climate and society are concealed simply because the period of historical observation is too limited. Moreover, identifying an area in which both internal and external causal factors appear to operate increases the interest and application of the case study. With these considerations in mind, various possible regions were reviewed and the Tigris-Euphrates Valley was selected (Figure 5.1). There are three reasons for this choice:

(a) it is best documented of all possible regions because the existence of clay tablets and the world's first writing system yields direct, although fragmentary, information about society back to 3000 BC;

(b) a major effort in reconstructing the population of the region provides a careful record of population dynamics for 6,000 years. Moreover, much of the spatial extent

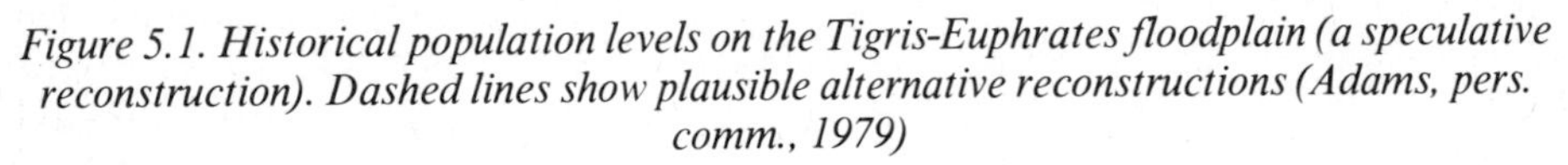

Figure 5.1. Historical population levels on the Tigris-Euphrates floodplain (a speculative reconstruction). Dashed lines show plausible alternative reconstructions (Adams, pers. comm., 1979)

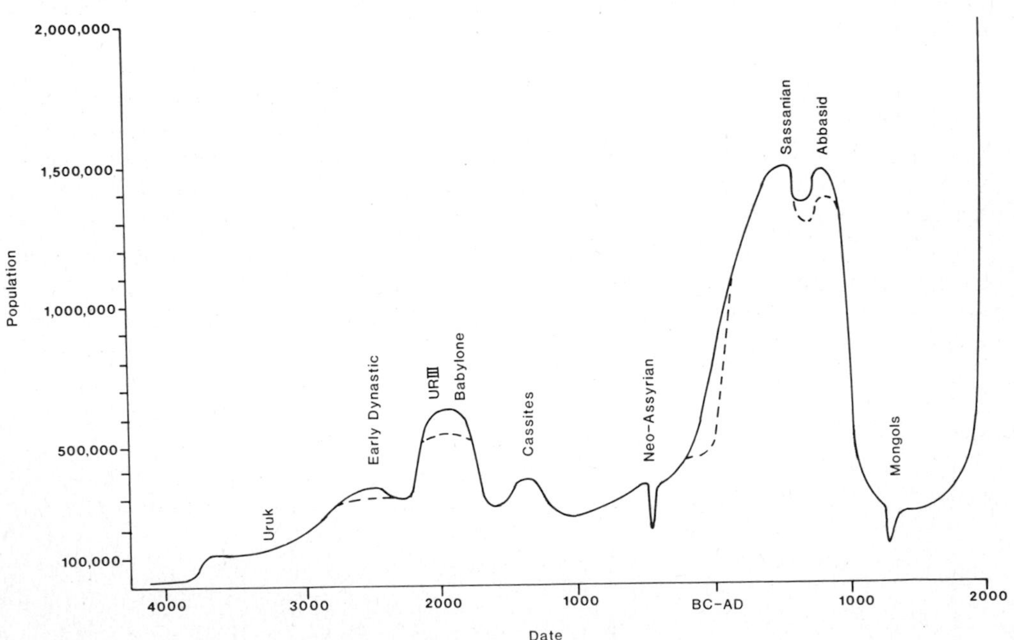

of the society is mapped; this reveals the centrality of irrigation, and thus of water supply and climate, to human life;

(c) sophisticated ecological analysis is central to the theories of archaeologists and anthropologists most knowledgeable about the region, and provides a rich base from which to construct hypotheses.

The decision to study this long historic period suggests that a simulation modelling approach, which emphasises the qualitative relationships between the major system components, would be useful. Such an approach makes it possible to bridge gaps in the historic record and to examine the long-term relationships of social and environmental components. Section 1 examines the pattern of population growth and the development of the irrigation system in the study area. In Section 2 we develop a simple model to illustrate the nature of the methodology and functional interrelationship of major system components. A more complex model is described in Section 3. Basic model runs are presented in Section 4 and discussed in Section 5.

1. POPULATION GROWTH AND IRRIGATION IN THE TIGRIS-EUPHRATES LOWLAND

Unlike most parts of the world, where reliable population data are a recent phenomenon, the floodplain of the Tigris and Euphrates rivers, ancient Mesopotamia, possesses a settlement record that can be reconstructed for six millenia. This historical record is the product of a literate, urban civilisation that was based on skilful manipulation of a sophisticated irrigation system. Much of the data base is archaeological in character, and is preserved as a result of the region's arid climate. Mud brick buildings and clay tablets have accumulated layer by layer as settlement has succeeded settlement on the same site, depositing a chronologically dateable network of refuse heaps into which archaeologists have delved for the last hundred years.

The result is a mass of information about society and polity. While more details are known about political events than about social and economic developments, the emerging data base permits a tentative reconstruction of the broad outlines of social history and demography. These reconstructions of population dynamics show that the Tigris and Euphrates lowland has experienced two and a half cycles of population growth and collapse. The population curves reconstructed by Adams (Figure 5.1) depict a series of oscillations of increasing amplitude that indicate decreasing system stability over time. This long-term record of increasingly large population fluctuations makes the Tigris and Euphrates a particularly interesting historical contrast to contemporary visions of exponential population growth.

Rainfall in the Tigris and Euphrates lowland is insufficient to provide a secure agricultural base, and the potential productivity of the fertile alluvial soils of the riverine lowland can only be realised by means of irrigation technology. This technology emerged slowly over the course of centuries as local groups gradually learned to manipulate the water resources available to them. Eventually, local leadership in organising local labour to carry out small-scale irrigation projects was supplemented by temple-based religious leaders who used their religious pre-eminence to coordinate secular activities. The temple storehouse emerged gradually as a central repository for surplus agricultural products. The development of a central storage function increased the importance of a specialised managerial population and its associated mercantile and military groups. These groups assumed a crucial coordinating function in the expansion of the irrigation system and the growth of the population. Over time, fully developed states emerged led by religious or kingly figures with a propensity for conspicuous consumption. This tendency was offset by the bureaucratic role played by the managers of the crucially important, increasingly integrated regional irrigation system (Walters, 1970).

Despite existence of a temporally long record, many lacunae exist. The survival of artifacts is spotty, events are imperfectly recorded, and much of the experience of everyday life is untransmitted to the present. Many events, significant in their cumulative effect but unexceptional individually, pass by unnoted. It is for these reasons that an effort to recreate comprehensively and understand a past society would be illusory.

Our effort is directed toward developing a better understanding of the qualitative aspects of the floodplain's irrigation civilisation. For both social and environmental processes and changes reliance must be placed on data at a scale of a century or longer. Direct physical evidence for the impact of shorter-term events is unavailable (Larsen *et al.,* 1978; Vita-Finzi, 1978), although the last several decades of instrumental records can be used as a basis for establishing the likely magnitude and recurrence interval of such extreme events as droughts and floods (Clawson *et al.,* 1971; al-Khashab, 1958; Rosenan, 1963; Ubell, 1971). Only political events can be dated with some confidence, and even for these there are difficulties (Bottero *et al.,* 1967), such as in the synchronisation of dynastic lists. It is not possible to specify precise quantitative measures for such variables as population growth rates, population size, and variation in stream flow. Rather, it is necessary to rely on fragments of literary data and environmental proxy information for much of our knowledge of events (Kay and Johnson, in press). For this reason, we have chosen a system dynamics type of modelling (discussed in Section 2) that emphasises the qualitative interaction of key variables.

There are a number of possible explanations for societal changes in the Tigris and Euphrates, some of which we have been able to examine explicitly. These explanations can be grouped into two broad categories: (a) those that rely on change derived from forces exogenous to Mesopotamia; and (b) those produced by internal causes. One possible external cause is climatic change. Evidence for climatic change in the last five millenia is scanty, and little direct evidence of the climate history of the Tigris and Euphrates has emerged from the lowland itself. The best data currently available is a recent review by Neumann and Sigrist (1978) of recorded barley harvest dates. This study indicates that harvesting began as much as thirty days later in 600-400 BC than it had a millenium previously. This combination of warmer, drier conditions might have encouraged over-irrigation and accounted for the first archaeologically identifiable case of salinisation. But for most of our climatic reconstruction, we must rely on environmental proxy data widely scattered across the highlands of Anatolia where the Tigris and Euphrates rivers have their headwaters.

Another possible external cause of population decline is invasion by neighbouring groups (Davis, 1949). Both organised states (e.g., Hittites c. 3500 BP) and tribal confederations (e.g., Gutians 4200 BP) episodically contributed to the destabilisation of Mesopotamian society. Often accompanied by the introduction of disease, frequently associated with internal disorganisation and civil unrest, invasions were a significant factor in Mesopotamian history.

Internal factors have also been touted as mechanisms causing collapse. Gibson (1974) has suggested that violation of fallow cycles would quickly lead to increased salinisation and a vicious cycle of lower yields, pressures for shorter fallow cycles, and ultimately collapse of the irrigation system. Since the basic agricultural regime is constructed around fallowing fields in alternate years in order to lower the water table below the plant root zone, any practice that reduced the period in fallow would encourage a rise in the water table. Once salts are no longer leached from the root zone, rapid loss of productivity takes place. The result is a progressive deterioration because declining yields encourage further shortening of the fallow cycle in order to replace the lost production. A crash inevitably ensues.

A second explanation argues for bureaucratic inefficiency as the primary explanation for collapse. In this scenario, leaders and bureaucrats either are ineffectual or excessively demanding. Leadership that can no longer organise labour to counteract siltation and extend the irrigation network, or a king so incompetent that revolt and civil war develop, could destroy the integration upon which the irrigation system depends. Alternately, rapacious demands for taxes to satisfy the aspirations of a privileged class could lead to destructive intensification practices on the part of the primary producers (Waines, 1977). With parts of the system no longer able to receive the water and logistic support upon which they have come to depend, extensive population collapse is initiated.

A third explanation stresses the role of conflict between settled and nomadic society. In this view, it is the conquest of settled folk by pastoral communities that leads to catastrophic collapse. These pastoralists may either be herders who normally exist on the margins of the irrigation system, linked to the settled communities symbiotically through patterns of seasonal grazing and trade, or coalitions of nomads both local and regional. Relations between the two groups have at times been so hostile that the irrigation areas have attempted to defend themselves by the Mesopotamian equivalent of a Great Wall dividing the desert from the sown (Barnett, 1963; Jacobsen, 1953). Any change in the power relationships of the two groups that reduces the ability of the settled population to defend itself effectively could result in conquest by a nomadic group which has only limited interest in maintaining an integrated irrigation system. Thus, nomadic conquest would result in population losses due to both combat and drastically reduced crop yields consequent on a disrupted and poorly maintained post-conquest irrigation network.

These explanations for population fluctuation are examined in IRRIG, our system dynamics simulation model of the Tigris-Euphrates irrigation society. In particular, the impact of climate fluctuations, especially drought, and fallow violation resulting in salinisation are the primary foci of the modelling effort. This exploration of the interaction of population dynamics, environment and technology begins with a simple two level model that establishes the basic relationships between components of the

Tigris and Euphrates irrigation system. Once these relationships are determined, the model is expanded in IRRIG to include a greater array of variables in a more realistic fashion. IRRIG then provides the framework within which we consider several of the major explanations for population change.

2. A SIMPLE TWO LEVEL MODEL

The cyclical growth and collapse of the Mesopotamian population is intriguing, and encourages us to seek a causal explanation. One way of understanding possible causes for this pattern is to develop a model in which all assumptions are made explicit. We adopt a system dynamics methodology which assumes that the persistent dynamic tendencies of a complex system arise from its internal causal structure rather than external disturbances of random events. In this section we summarise the basic ideas of system dynamics[2] and apply them to a simple two level model.

From the point of view of system structure, systems with the same structure exhibit the same behaviour patterns. Since a number of well-known ecological models yield cyclical growth and collapse, it is instructive to study first a simple model which yields this qualitative behaviour. Consider a system for which the total population and the carrying capacity are the two state variables or "levels". The carrying capacity need be only loosely defined as the ability of the society to sustain a corresponding level of population. This definition differs from that normally employed in ecology, where carrying capacity is defined as the ability of the physical environment to sustain a population. Viewed from a more social perspective, ecological carrying capacity is redefined whenever technology devises new ways of deriving support for human populations. In this context, carrying capacity aggregates quantities such as the size of the irrigation system, the amount of available water, the level of stored food, and the organisational abilities of society.

One possible, inherently Malthusian, interpretation of the observed behaviour pattern in the Mesopotamian population is that at low levels of population the society is able to increase carrying capacity at a rate faster than population growth. However, at higher levels of population, growth in the carrying capacity can no longer keep up

2. The central concept used in system dynamics is the principle of feedback. Feedback exists whenever there is a closed chain of causal relationships. In a positive feedback process a change in a variable leads to further change and produces exponential growth or collapse. A negative feedback process counteracts a change, and moves the system toward equilibrium or a specific goal. System dynamics models link together many such positive and negative feedback loops. Implicitly associated with this feedback structure are time delays, nonlinear relationships that are also assumed to be important in determining system behaviour. A nonlinear relationship causes the feedback loop to vary in strength depending on the state of the system. Hence, different combinations of feedback loops might be dominant and shift the system's behaviour. A model composed of several feedback loops that respond to each other nonlinearly, and with significant time delays, can exhibit a wide variety of complex behaviour patterns. Since the consequences of such relationships are difficult to understand analytically or intuitively, it is convenient to study such models using computer simulation. Although system dynamics models are frequently written in Dynamo, they can also be written in a general purpose language such as Fortran or Basic.

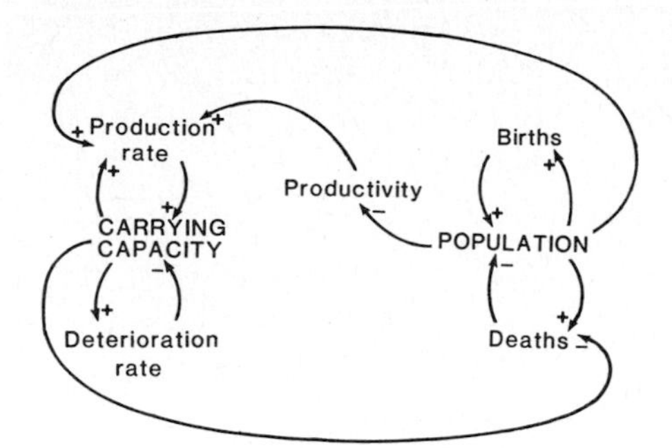

Figure 5.2. Causal loop diagram of simple ecological model

with the growth in population. It then might be conjectured that the society cannot sustain this high level of population and the population falls to a much lower level from which it eventually recovers and begins another cycle.

Possible causal feedback relationships which are consistent with the above description of Mesopotamian society are shown in Figure 5.2. These relationships include positive and negative feedback loops relating births, deaths and population. If the positive feedback loop dominates, the population exhibits exponential growth. The level of population is linked to the carrying capacity by the latter's influence on the number of deaths. Carrying capacity is determined by its production and deterioration rate. For simplicity, we assume that the production rate is proportional to the population. Productivity (increment in carrying capacity per person input) is assumed to decrease with increasing population, and the carrying capacity deterioration rate is assumed to be a constant fraction of the carrying capacity.

A computer simulation model was constructed to make these qualitative causal relationships more explicit. Normal conditions are arbitrarily assumed to correspond to the ratio of population to carrying capacity being near unity. Typical behaviour of the population and carrying capacity is shown in Figure 5.3a. Inspection shows that the growth in the carrying capacity initially exceeds that of the population and that both increase exponentially. However, after about 200 years, the population exceeds the carrying capacity and then falls to a sustainable level. This qualitative behaviour mode, similar to sigmoidal or logistic growth, arises from the linked positive and negative feedback loops that respond to each other nonlinearly, but with no significant time delays.

To understand the observed oscillatory population behaviour mode, we look for the important time delays in the system. Time delays are ubiquitous in dynamic systems and play an important role in stabilizing or destabilizing the system. For our example, it is easy to identify time delays between a change in the carrying capacity and the number of deaths, and a change in the population and the productivity of the resource base. If we assumed that these delays are on the order of ten and twenty-five years respectively, the population and carrying capacity do exhibit oscillatory behaviour as shown in Figure 5.4a. Note that the period of the oscillations (the time between relative

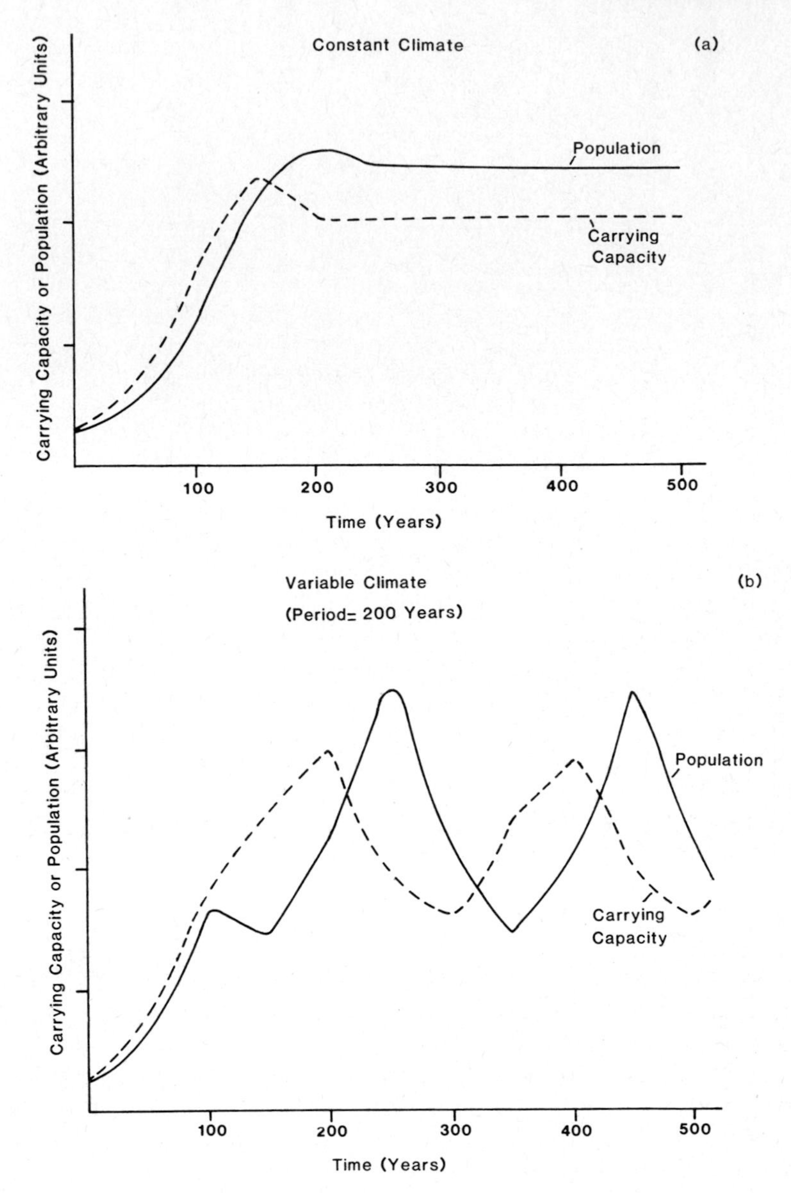

Figure 5.3. Behaviour of the population (solid line) and carrying capacity (dashed line) in ecological model with no time delays: (a) constant climate, (b) variable climate

maxima or minima of the population) is approximately 200 years for the particular values chosen for the numerical constants in the model.

Since we are interested in the effect of climatic fluctuations on society, it is instructive to study the response of our two models to climatic fluctuations. We

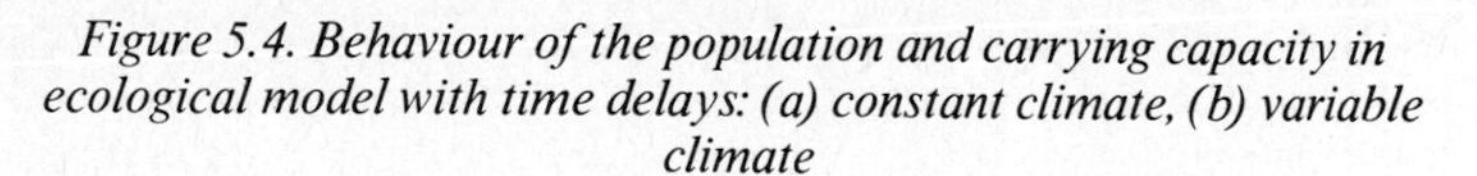

Figure 5.4. Behaviour of the population and carrying capacity in ecological model with time delays: (a) constant climate, (b) variable climate

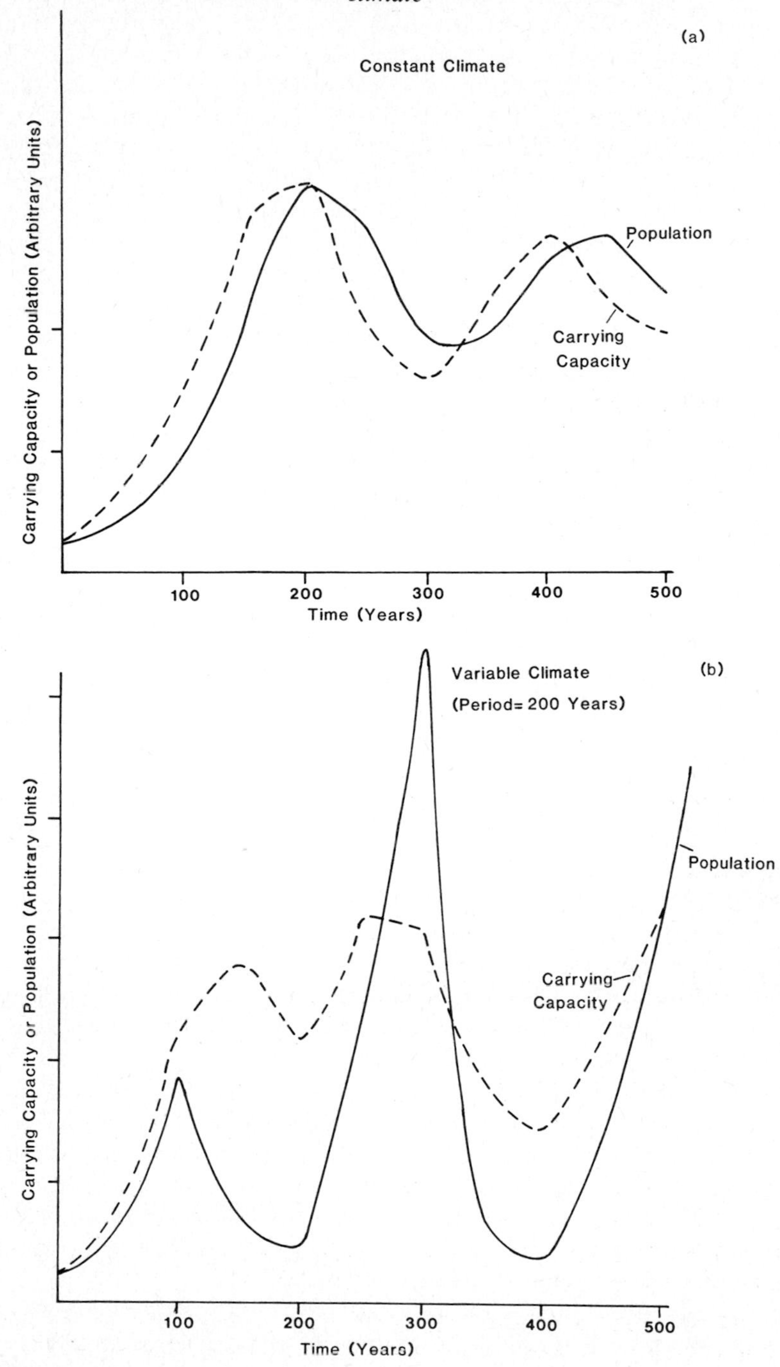

associate climatic fluctuations with a variable streamflow and hence with an externally imposed variation in the carrying capacity. This variation is in addition to the changes in the carrying capacity due to the internal dynamics of the system. Let us assume that streamflow varies sinusoidally, that is, it is characterised by its period and amplitude, and that the amplitude of the riverflow variation is such that it changes the carrying capacity by a maximum of 50 per cent. (The form of the streamflow variation and the magnitude of its amplitude are chosen only for illustrative purposes.) Taking the period of the streamflow to be 200 years, and assuming no time delays, Figure 5.3b shows that the population and carrying capacity oscillate with the same period as the externally imposed climatic fluctuations and that their maxima coincide with the peaks in riverflow. If we reduce the climatic periodicity to 25 years, the population and carrying capacity exhibit unperturbed behaviour modes.

A variable riverflow of 200 year periodicity imposed on the system with time delays, yields the oscillatory population and carrying capacity behaviour shown in Figure 5.4b. Note that population maxima do not coincide with peaks in riverflow. If the riverflow period is reduced to 25 years, there is little effect on the system.

One general observation drawn from these results is that the cause of the observed fluctuations in a state variable such as population may be internal or external to the system. Since frequently the cause is a combination of the two, the peaks in the population do not necessarily coincide with peaks in the external forcing variable. Changes in the streamflow that occurred over a 200 year period were reflected in significant changes in the population. However, changes in climate periodicity that occurred over a much shorter time scale had no significant long-term effect. Such a conclusion is valid in general as long as the external perturbations are small enough in magnitude so as not to cause major changes in system structure.

3. THE TIGRIS AND EUPHRATES MODEL

Our initial success with a simple model encourages us to develop more sophisticated models of an irrigation-based society. In this model (IIRIG), we place our emphasis on internal factors that promote change. The only external causal factor explicitly considered is climatic variation and change. Treating the Tigris and Euphrates as a closed political and economic system is an unrealistic simplification of historical reality. Trade and tribute from non-irrigated areas provided important resources to the sustenance of Mesopotamian life in many epochs. Denial of access to those resources could have serious implications for the stability of the central government, certainly a factor in the long decline of the Abassid Caliphate before its destruction by the Mongols in AD 1258. Similarly, invasions by ephemeral nomadic confederations or by hostile states could, and did, contribute to the disruption of Mesopotamian irrigation systems and the decline of the population dependent upon them. While recognising the role played by such factors, we hold their specific exploration in abeyance for future iterations of the model.

As a first step in developing IRRIG, our model of the Tigris and Euphrates irrigation system, we show some of the basic feed-back loops in Figure 5.5. Most of these are self-

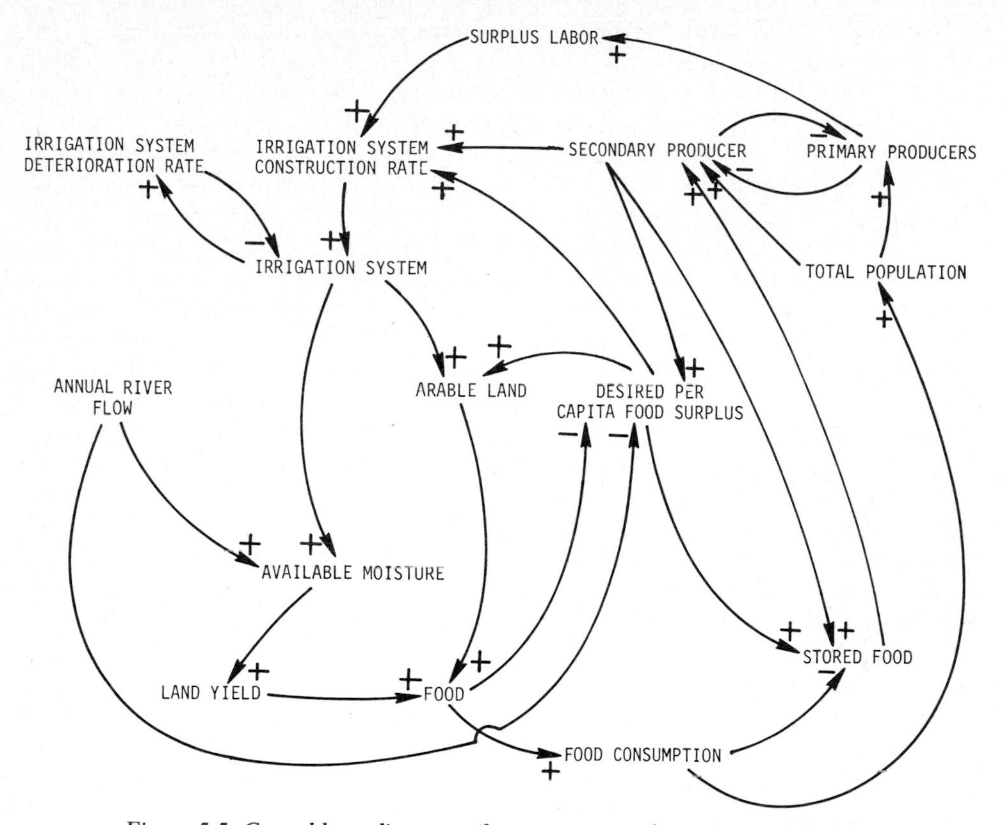

Figure 5.5. Causal loop diagram of IRRIG*, a system dynamics model of the Tigris-Euphrates irrigation scheme*

explanatory at this qualitative level. The important state variables are population, size of the irrigation system, level of stored food, and the size of the harvest. The latter is affected in part by the available moisture for agriculture and the salinity of the soil. Since the causal loop diagram of Figure 5.5 only includes some of the qualitative relationships, we give in the following paragraphs a more detailed outline of IRRIG. A discussion of the limitations of IRRIG is reserved for Section 5.

The total population is divided into two groups, primary producers who are mainly involved in agriculture, and secondary producers who are engaged in trade, irrigation system organisation, extension and maintenance, the performance of religious functions, specialised artisanal enterprises, and military activities. The irrigation system can exist in either of two states: (a) *disaggregated,* characterised by a discrete series of locally managed units localised along meandering natural stream channels, each unit comprising a small total population; and (b) *aggregated,* where artificially constructed, centrally managed canals integrate local cultivation units into a larger system and larger population totals can be maintained. As indicated below, this division of the population and the irrigation system into two basic types is useful in examining the mechanisms involved in irrigation system expansion.

Two of our fundamental assumptions are that both population growth and the need for a stored food reserve (due to the ever-present danger of variation in available moisture) are mechanisms that promote construction of a more sophisticated and integrated irrigation system. The assumption of these two mechanisms makes it convenient to conceptually separate the total irrigation system into locally- and centrally-managed components. The former is extended and maintained by the primary producers who extend the local irrigation system at a rate that keeps population growth and available food supply in a rough balance. That is, the primary producers increase the irrigated area in small increments at the margin of existing cultivation. Such increases are small in scale, and locally available labour, organised and controlled consensually by local leadership, is sufficient to extend and maintain the local irrigation system independent of the size and of the existence of the secondary producers.

When the expansion of the local irrigation system approaches its technological and managerial limits, there is increasing incentive for the growth of a specialised group in the population. These secondary producers (and the other specialised groups who serve and/or are associated with them) organise and centrally manage an expanded irrigation system. In the Tigris and Euphrates, this specialised elite developed in urban settlements where priests, the servants of the major local diety, performed mediatory and organisational functions that transcended the boundaries of local kin groups. In IRRIG the strength of this central authority at any point in time is represented by the ratio of specialised to primary producers in the total population.

A paramount function of this specialised elite was the development of stored food reserves that buffered the community from the vicissitudes of short-term environmental fluctuation. The existence of such reserves also served as an attractive magnet for peripheral populations whose movement toward the better-endowed centre augmented urban population growth and increased the demand for larger and more secure food supplies. Although only a small proportion of this migration flow would cease to be primary producers, it is by means of such spatial redistribution of population as well as social and economic change that primary producers can be transmuted into secondary and specialised producers or, under altered circumstances, *vice versa*.

In IRRIG, it is assumed that the main goal of the secondary producers is to expand the central irrigation system in order to keep the stored food reserve above a desired minimum level. Their ability to do this is determined by the percentage of specialised producers in society, and by the amount of surplus food already available. The central bureaucracy also has the option of investing the surplus food in more self-indulgent uses rather than in increased food consumption, surplus storage and the extension of the irrigation system.

Feedback between the population and the irrigation system loops is positive since an increase in the irrigation system construction rate due to population growth expands the available cultivated land, makes it possible to tap a greater percentage of water from the main rivers, and hence to increase yields, the size of the harvest and the level of stored food. These increases lead to further increases in population which further spur the rate of irrigation system construction.

In addition to these positive feedback loops, there are several important negative feedback loops in the system. An increase in available moisture can, in the absence of careful control of its application at the field level, set in motion a rise in the water table and an increase in salinity. A substantial increase in soil salinity has a negative impact on crop yields, an environmental management problem well documented in ancient Iraq (Jacobsen and Adams 1958). A decrease in the amount of stored food can also have a negative impact on yields by encouraging violation of fallow (Gibson, 1974). Farmers in the Tigris and Euphrates traditionally have followed an alternate year fallowing pattern that depends on deep-rooted shrubs acting as natural tube wells to draw down the water table. Should prolonged drought reduce the stored food supply below the desired minimum, or should rapacious taxation affect the same result, intensified cropping that kept fields in production for longer periods before being fallowed would be the farmer's only alternative. Higher salinity levels and lower yields would be the inevitable consequence of such practices. These degradational tendencies are likely to appear first in the older, more established components of the irrigation system that have been in production longest. This process of degradation in core areas could contribute both to the search for new technological solutions to water management and to the extension of the irrigation system at its margins into new areas. The consequence of shortened fallow cycles would be the development of a treadmill effect analogous to that afflicting many irrigation systems today (Worthington 1977; White, 1978), where expansionary gains at the margins of the system are offset by losses in the older core areas.

Increases in the amount of cultivated land due to irrigation system expansion not only increase food supply, but they also increase land development costs. Because the best and most easily developed land will be brought into production initially, the cost of developing new land at greater distances from the source river will rise. Such cost acts as a brake on the irrigation system construction rate. Other constraints also act to slow construction. One is the greater labour costs required to dig longer canals from river to undeveloped land with consequent increases in the drain on stored food needed to pay for such labour. Another constraint is the finite amount of water that can be extracted with existing technology from the available water supply. Because we base IRRIG on conditions characteristic of the first cycle of population growth, we hold constant the level of technology available. Subsequent epochs did develop new technologies for increasing the amount of water extracted from the river system, and this represented an important condition for later growth cycles.

Also important is the increased proportion of available surplus stored food that must be channelled from construction into maintenance of the existing canal infrastructure as the irrigation system grows. The model assumes that the deterioration rate can be controlled as long as central authority remains intact and population density remains high enough to provide sufficient labour. It is also clear that the larger the system the more maintenance expenditures become competitive with, and have a negative impact on, the irrigation system construction rate (Adams, 1974).

Since rainfall in the alluvial lowland is insufficient for agriculture without irrigation, streamflow is based on precipitation occurring in the adjacent highlands. Streamflow can be regarded as a surrogate for climatic fluctuation, since any variation in regional climate will produce a concomitant change in available moisture for irrigation. The twenty-five years of instrumental record do not encourage confidence that the full range of streamflow variability is contained in existing data. It is possible to incorporate random variation about the mean into the model to simulate the effect of extreme events. The model also permits testing of the impact of short-term but more persistent variation.

4. MODEL RESULTS

We describe in this section some typical runs of our computer model IRRIG. Because of the assumed finite resource base of the system, an important dynamical question is whether the population will be able to reach a final equilibrium state of maximum size or collapse after a period of initial growth. Of particular interest is the behaviour of IRRIG under constant and variable climate conditions.

Constant climate

The total amount of water flowing in the river per year or the "water supply" is the surrogate for climate in IRRIG. As one test of the relationships in IRRIG, we made a "standard" run for which the water supply does not fluctuate. An established mean streamflow of 6.2×10^9 cubic metres per year is employed, a figure that is consistent with the known instrumental record (al-Khashab, 1958; Ubell, 1971). The behaviour of the total population is shown in the solid line in Figure 5.6. The population and the total irrigation system (not shown) exhibit similar behaviour, and both achieve equilibrium after an approximately one thousand year period of sustained growth. Yields are limited only by the amount of water available for irrigation and not by salinisation. Further growth in the population is constrained by the finite resource base rather than by salinisation or some other internal cause. The model society is able to support a large population by lowering its per capita consumption by approximately 20 per cent from the desired amount. The irrigation system is overextended and hence less than the optimum amount of water is delivered to the fields. Although fallow is violated, salinisation does not develop since the fields are underirrigated. The organisational structure of the society shifts increasingly away from local control to centralised management of irrigation system construction and maintenance.

Variable climate

Several runs of IRRIG were made with a variable water supply which is assumed to vary randomly about its mean with a persistence time of approximately 50 years. The mean value of the water supply is equal to that assumed in the constant water supply run described above; the standard deviation from the mean is assumed to be thirty per cent of the mean. These values for persistence time and standard deviation are taken for

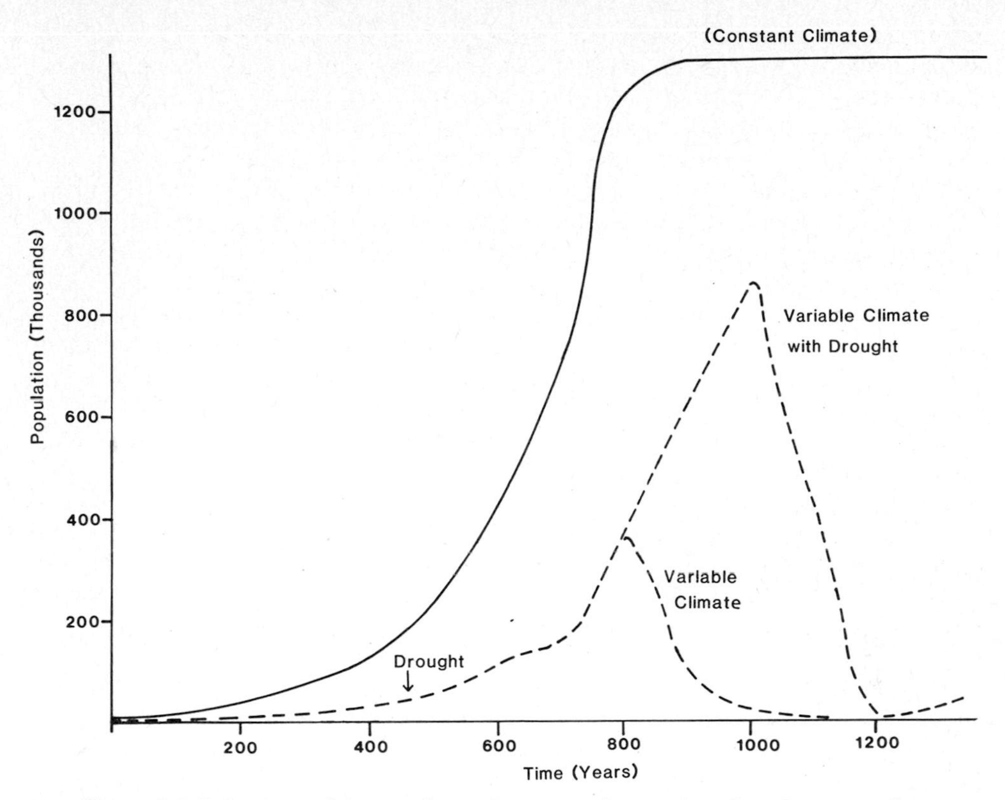

Figure 5.6. Behaviour of the population in IRRIG *under varying climatic assumptions*

illustrative purposes only, since the period of instrumental record is too short to produce meaningful data.

Two representative examples of the behaviour of the population for different variable climates are shown in Figure 5.6 (dotted lines). In the "non-drought stressed" case the climate varies as discussed above. In contrast, the water supply in the "drought stressed" example has a ten-year drought imposed at year 500 in addition to the fluctuations in the water supply that occur in the non-drought stressed case. Hence the differences in the water supply between the two examples occur during the years 500-510. The water supply is at least 30 per cent below the mean in all ten years of the drought stressed run, but is below the mean in only seven out of the ten years of the non-drought stressed run. Note that the drought occurs during a period of population growth.

The behaviour of the population in both variable climatic runs (Figure 5.6) is similar — a period of growth followed by a rapid collapse. The behaviour of the irrigation system, which is now shown in the figures, is similar. The population and irrigation system behaviour mode of growth and collapse for variable climate is in contrast to the sigmoidal behaviour mode obtained for constant climate. Clearly the addition of variability in the water supply is enough to change the system's behaviour dramatically. In addition to the different behaviour modes, the rate of initial growth of the population is, as might be expected, less for the variable climate than the constant climate runs.

Although the population does exhibit similar qualitative behaviour in both variable climate runs, the population does show a longer period of sustained growth in the drought stressed case. Population in the drought stressed run rises to a maximum of 960,000 people at year 1025 rather than the maximum of 270,000 people at year 800. The explanation for this result is that the drought was imposed at a time when sufficient food resources were available. Thus, although the food resources were lowered by a factor of two in the following twenty-five year period in order to maintain adequate consumption, there were sufficient reserves remaining to allow the central bureaucracy to expand the irrigation system and to replenish the depleted food resources. At year 800 when the non-drought stressed population shown in Figure 5.6 begins to collapse, the population of the drought stressed run has 50 per cent greater food resources and is better able to sustain itself.

We now determine the specific scenario for the collapse of the population and the associated infrastructure. As can be seen from Figure 5.6, the non-drought stressed society at year 800 had a population maximum of 370,000 people, a population much below the resource-limited maximum of one million people assumed in the model. Not shown in Figure 5.6 is the fact that at this time there were approximately 2.5 years equivalent of stored food, although the stored food resources had decreased slightly for the previous 100 years. The water supply is slightly greater than normal during the 200 years before year 800 and dips to slightly below normal during the succeeding 25 years. This slight change is not sufficient to cause a collapse in the population by itself. However, the decrease in the water supply is sufficient to cause the food resources to decrease further and to cause the strong central bureaucracy to extend the central irrigation system (now about 80 per cent of the total system) to maintain reserves. This extension of the irrigation system lowers the food reserves even further and causes the primary producers to violate fallow. Unfortunately, this violation of fallow leads to rapidly developing salinity, decreased yields, and a destitute population. The central bureaucracy collapses and the central irrigation system is no longer maintained properly. The collapse of the central irrigation system leads to reduced yields and the downward trend becomes greater still.

Since the drought stressed society had greatly expanded its irrigation system during the imposed drought of years 500 to 510, it had greater resources during year 800 and thus was not vulnerable to collapse. The scenario for the collapse of the population in Figure 5.7 beginning at year 1025 differs from that given above for the non-drought case. In this case the population of 960,000 people is at its assumed resource limit. The food reserves were now reduced to less than one year and yields were low because of the inefficiency of the irrigation system at its technological limit. In this case the cause of the collapse is more direct. That is, a drop in the water supply to about 75 per cent of normal causes a decreased yield, starvation, and a collapse of the central bureaucracy. Collapse of the central bureaucracy then leads to a lack of maintenance in the central irrigation system. Salinisation is never a factor since the society is short of water.

Several other runs of IRRIG were made for other assumed climate conditions. For example, a ten-year drought was imposed at year 850 in addition to the fluctuation in the water supply discussed above. As can be seen from Figure 5.6, this drought occurred during a time of population collapse. Since the immediate cause of the collapse was reduced yields due to salinity, the short-term response was a pause in the collapse of the population. That is, less water for irrigation led to reduced salinity. This pause was only temporary and the population then continued its collapse. Other runs were made with a reduced mean water supply. As expected, the initial growth in the population decreased as the mean was reduced until a point was reached at which the population remained constant.

We emphasise that the above scenarios are only qualitative in nature. In all of the runs the model exhibited a high degree of instability, e.g., the rates of growth and collapse were more rapid than might be expected from our discussion in Section 3. This behaviour is in part due to our decision in the analysis to exaggerate the importance of various causal mechanisms. However, the behaviour modes of the model are sufficiently analogous to those observed historically to give some confidence that the model's qualitative relationships are reasonable.

5. DISCUSSION

Two general conclusions emerge from IRRIG. The first insight that the model offers is that, within the constraint of a fixed resource base, climate variability leads to collapse, whereas climate constancy encourages an equilibrium state. A stable equilibrium results when there is a fixed resource base where all other factors are held constant. The constant climatic run contains the same negative feedback loops as do the variable climatic runs. In the constant climatic scenario, positive and negative feedback loops are in balance and population approaches equilibrium with only a light overshoot. However, the system is finely balanced. Once climatic variability is added to the array of existing negative stresses, the achievement of population stability proves an illusory objective. Instead, climatic variability enhances the intensity of negative feedbacks and increases their impact on society. The result is a population oscillation in which time delays retard but cannot remove the impact of negative environmental fluctuations.

A second conclusion that can be drawn from IRRIG is that extreme climatic events can stave off or encourage population collapse. The key variable is the timing of the extreme event in relationship to societal well-being. Beneficial consequences, e.g., enhanced social stability and population growth, result when drought and adequate stored food reserves occur simultaneously. If these conditions also coincide with a period of general population growth and environmental prosperity, then the positive consequences are enhanced. The possession of sufficient surplus stored food reserves enables society to respond to drought by extending the irrigation system. Stressful events and existing coping strategy allow society to attain a population total greater than in other variable climatic runs in which no additional drought was imposed. The reason for this positive effect lies in the stimulus provide to increased infrastructure development. The new irrigation system additions better prepared the society to cope with future reverses by increasing total yields and by enhancing stored food capacity.

Nonetheless, the long-term behaviour of the variable climate system under both drought stressed and non-drought stressed conditions remains the same. A long period of population growth and irrigation system expansion is followed by a subsequent collapse. In a long-time series the disappearance of short-term drought impacts is not surprising; they are simply absorbed into the larger patterns of events. This does not mean that such short-term events are unimportant. On the contrary, they may have a significant impact on society if they happen to occur during times of low food reserves, an over-extension of resources, and bureaucratic decline. Only rarely does the broad outline of these short-term events appear in the historical record, and they are not specifically tested in the present IRRIG runs.

The positive short-term response to drought described above occurs in part because of the explicit assumption in IRRIG that a goal of society is to maintain stored food reserves at some fixed level. Such a goal is desirable whenever there is a past history of food supply shortfalls. That is, the incentive for the establishment of an appropriate level of stored food reserve also depends on the presumptive demands of the privileged segment of society (the secondary producers). These demands give secondary producers preferential access to the food reserves that they help to create and that they largely control. Thus, the inclusion of a fixed rather than a variable goal for desired stored-food reserves omits an important dynamical aspect of society. The assumption of a fixed goal is made in order to simplify the analysis and should be modified in future work.

IRRIG does not include the option of incorporating resources from outside its geographical area. These resources often were very important to floodplain society. Deficient in timber, stone, minerals and precious metals, the riverine states produced both finished goods (tools, jewellery, cloth, etc.) and agriculture commodities (dates, rice, etc.) that were unattainable in the cooler and moister neighbouring uplands. Patterns of trade, raid and tribute emerged to provide access to these desired commodities. The investment of resources obtained (often by forced tribute or taxation) from outside the floodplain was an important feature of lowland political economy that is not included in the structure of the IRRIG model. This restriction on the resource base in IRRIG also sets an upper limit to the total amount of water that is available to the irrigation system from the floodplain rivers. The irrigation technology is also assumed to be constant, and this limits the percentage of available water that can actually be extracted by the irrigation system. The introduction of technological innovations, such as lift devices, would increase the water-use efficiency of the irrigation system and improve its ability to cope with fluctuations in available moisture. However, the absence of a provision for technological innovation is reasonable for the first cycle of population growth and decline that is represented by the IRRIG model.

Restrictions on the resource base and the irrigation technology imply that there is a finite amount of water and hence an upper limit to the size of the population that can be supported by the irrigation system. These limits increase the vulnerability of the society at a point when these limits are reached. Historically, irrigation civilisations in the Tigris and Euphrates lowland were able to reduce their vulnerability to environmental risk by gaining access to outside resources by trading activities or by military conquest. The result of this integration was that many floodplain societies reduced their vulnerability by the buffering effect of stored reserves accumulated from outside the territorial limits of the irrigation system itself. In this sense, the increased security of lowland society was purchased by the export of that risk onto surrounding populations. The absence of these options in IRRIG detracts from its realism; increased agricultural intensification on existing arable land is the only mechanism which the model employs.

At a scale of centuries and millenia, it is not possible to discern specific incremental changes in vulnerability to disruptive, recurrent climatic events, like drought. Only broad relationships, as they become evident over long periods of time, can be established. However, we can assume that particular mechanisms for lessening societal vulnerability have been adopted, as, for example, through increased food reserves, expansion of the irrigation system, centralisation of irrigation system control, and innovations in water management and distribution technology.

In IRRIG, two adaptive mechanisms are incorporated. First, we assume that there is a fixed goal of accumulation of stored food or surplus to provide the community with protection against frequent seasonal or annual shortfalls. In particular, we assume that food storage is the major societal buffer against recurrent drought and flood events and depends on the past experience of the society. The second adaptive mechanism is the establishment of a centrally organized irrigation system. Although total destruction of all of the small local irrigation systems along the river would be unlikely, any one local irrigation system would be vulnerable to total collapse. Linking together small irrigation units has short-term adaptive value and leads to a more regular water supply and better yields. The result is a positive feedback loop in which the greater surplus generated by the irrigation system provides additional resources. These resources are invested in further irrigation system expansion by supporting the central bureaucracy and the labour costs of maintenance and construction. This expansion of the local resource base, and its integration into a larger, more hierarchical organisation, is supported by the available archaeological and historical record. As discussed in Section 4, the model runs do indeed demonstrate the adaptive value of this mechanism: the expansion of the irrigation system provides the resources with which to support continued population growth over a sustained time in spite of a variable food supply.

These relationships raise the issue of whether or not the increased integration and organisation required by the adaptive mechanisms lead to greater societal vulnerability to large magnitude, infrequent events. For example, once the irrigation-based society develops an integrated water management system, changes in the upstream areas have enhanced impacts on downstream groups. Thus, severe climatic-related impacts in one area might produce serious adverse consequences elsewhere in the region. Do mechanisms which insulate the society from relatively frequent climatic, political, or economic disturbances imply a greater potential for catastrophe from rare, extreme events in the future?

Once stress in the form of a variable climate is introduced in IRRIG, society responds to mitigate its effects. Better and more integrated water management produces higher yields and food surpluses which support a larger population. This larger population in turn provides the labour to enhance system productivity and efficiency. For nearly one thousand years the overall population trend is progressively increasing. These adaptive strategies prove to be counter-productive, for under variable climate conditions collapse ultimately occurs. However, further work needs to be done to identify which adaptive strategies are most likely to contribute to catastrophic collapse. For instance, the relative impact of a ten-year drought on a centralised as opposed to a decentralised society could be determined. The relative merits of a large or a small food reserve could be assessed under conditions of extreme drought. By working out such scenarios one-by-one, both the mechanisms which lessen vulnerability, and their relationships to the potential for catastrophe, could be explored.

Climatic variation by itself is not the sole cause of catastrophe, since in the model runs the imposition of an earlier, intense ten-year drought had no effect on the population. But an external forcing variable such as climate can have a catastrophic impact on society during a period of internal stress. A cascade of misfortune (Post, 1977) then sets in. A scenario of the process might be as follows. The system created to cope with the anticipatable shortfall in available moisture, proves incapable of dealing with a new constellation of events. A rapid shift to a new system with smaller population, localised administration, less stored food, and greater vulnerability to recurrent environmental impacts emerges. This decentralised system is able to survive extreme events at the cost of the heightened vulnerability of individual components to more frequent occurrences. This benefit is insufficient to prevent future generations from attempting once again to pursue the remembered golden epoch of increased security from everyday hazards that the integrated, centrally managed irrigation system represents.

Additional questions are also generated by the development of the IRRIG model. A question that merits further consideration concerns the primary motivations for the development of the irrigation system. Far more theories have been offered to explain collapse than have been developed to elucidate the initial growth of an integrated irrigation system. Present explanations (Redman, 1978) stress multiple feedback processes associated with the rise of urbanisation, the beginnings of the state, and the inception of social stratification, but specific causal mechanisms remain obscure.

What is the role of non-irrigation based resources in buffering the floodplain irrigation system? To a certain extent this question is a consequence of assuming that the lowland floodplain was a closed system. The taxes generated from peripheral provinces, the products essential to the floodplain economy derived by trade, and the booty brought home by conquering warriors were not accounted for in IRRIG. Neither the role of animals as food storage buffers in the sedentary community, nor the complex exchanges of people and animals between nomadic pastoralists and irrigation farmers are explicitly acknowledged by IRRIG. It is possible that the resistance of the integrated irrigation system to catastrophic shock would be greatly enhanced by the inclusion of such factors.

Finally, is there a relationship between climatic variability and the size of a society's stored food reserves? Does greater variability encourage provision for greater stored food supplies, either in central granaries or on the hoof? When does such storage become counterproductive, placing excessive stress on the production system and the environment that sustains it? The answers are unclear.

It is evident from the above discussion that the mechanisms for collapse in complex systems are not unicausal and differ depending on various factors. Hence it is not surprising that there are a number of differing theories for collapse. All may be correct in the sense that they may be relevant at one time or another. The problem is to determine which theory, or combination of theories, is correct for which society at what time.

Note: This chapter was completed without the benefits of new insights into the patterns of human settlement and land use in Mesopotamia recently published in *Heartland of Cities* by Robert McC. Adams, (1981) University of Chicago Press, Chicago.

REFERENCES

Adams, Robert McC. (1965) *Land Behind Baghdad: A History of Settlement on the Diyala Plains.* Chicago: University of Chicago Press.

Adams, Robert McC. (1974) "Historic Patterns of Mesopotamian Agriculture," in Theodore E. Downing and McGuire Gibson (Editors), *Irrigation's Impact on Society,* Tucson: University of Arizona Press, pp. 1-6.

Barnett, R.D. (1963) "Zenophone and the Wall of Media," *Journal of Hellenistic Studies,* Vol. 83, pp. 1-26.

Bottéro, Jean *et al.* (1967) *The Near East: The Early Civilisations.* Trans. by R.F. Tannenbaum. New York: Delacorte.

Bryson, Reid A. and Murray, Thomas J. (1977) *Climates of Hunger: Mankind and the World's Changing Weather.* Madison: University of Wisconsin Press.

Clawson, Marion, Landsberg, Hans H. and Alexander, Lyle T. (1971) *The Agricultural Potential of the Middle East.* New York: American Elsevier.

Davis, William Stearns. (1949) *A Short History of the Near East from the Founding of Constantinople (330 A.D. to 1922).* New York: Macmillan.

Dols, Michael W. (1977) *The Black Death in the Middle East.* Princeton: Princeton University Press.

Gibson, McGuire. (1974) "Violation of Fallow and Engineered Disaster in Mesopotamian Civilization," in Theodore E. Downing and McGuire Gibson (Editors), *Irrigation's Impact on Society.* Tucson: University of Arizona Press, pp. 7-19.

Jacobsen, Thorkild. (1953) "The Reign of Ibbi-suen," *Journal of Cuneiform Studies,* Vol. 7, pp. 36-47.

Jacobsen, Thorkild and Adams, Robert M. (1958) "Salt and Silt in Ancient Mesopotamian Agriculture," *Science,* CXXVIII, No. 3334 (21 November), pp. 1251-1258.

Kay, Paul A. and Johnson, Douglas L. (1981) "Estimation of Tigris — Euphrates Streamflow from Regional Paleoenvironmental Proxy Data," *Climatic Change,* III, pp. 251-63.

al-Khashab, Wafiq Hussein, (1958) *The Water Budget of the Tigris and Euphrates Basin.* University of Chicago, Department of Geography, Research Paper No. 54.

Larsen, Curtis E. and Evans, Graham. (1978) "The Halocene Geological History of the Tigris — Euphrates — Karun Delta," in William C. Brice (Editor), *The Environmental History of the Near and Middle East Since the Last Ice Age.* New York: Academic Press, pp. 227-244.

McNeill, William H. (1963) *The Rise of the West: A History of the Human Community.* Chicago and London: The University of Chicago Press.

McNeill, William H. (1977) *Plagues and Peoples.* New York: Doubleday.

Neumann, J. and Sigrist, R.M. (1978) "Harvest Dates in Ancient Mesopotamia as Possible Indicators of Climatic Variations," *Climatic Change,* Vol. 1, pp. 239-252.

Post, John Dexter. (1977) *The Last Great Subsistence Crisis in the Western World.* Baltimore: John Hopkins University Press.

Redman, Charles L. (1978) *The Rise of Civilisation: From Early Farmers to Urban Society in the Ancient Near East.* San Francisco: W.H. Freeman.

Rosenan, N. (1963) "Climatic Fluctuations in the Middle East during the Period of Instrumental Record," in *Changes of Climate: Proceedings of the Rome Symposium.* Arid Zone Research No. 20. Paris: UNESCO, pp. 67-73.

Ubell, K. (1971) "Iraq's Water Resources," *Nature and Resources,* VII, No. 2, pp. 3-9.

Vita-Finzi, Claudio (1978) "Recent Alluvial History in the Catchment of the Arabo-Persian Gulf," in W. C. Brice (Editor), *The Environmental History of the Near and Middle East Since the Last Ice Age.* New York: Academic Press, pp. 255-261.

Waines, David (1977) "The Third Century Internal Crisis of the Abbasids," *Journal of the Economic and Social History of the Orient,* Vol. 20, pp. 282-306.

Walters, Stanley D. (1970) *Water for Larsa: An Old Babylonian Archive Dealing wth Irrigation.* New Haven and London: Yale University Press.

Wheeler, R. E. Mortimer (1968) *The Indian Civilisation: Supplementary Volume to the Cambridge History of India.* (Third edition). Cambridge: Cambridge University Press.

White, G. F., (ed.) (1978) *Environmental Effects of Arid Land Irrigation in Developing Countries.* MAB Technical Notes 8. Paris: UNESCO.

Worthington, E. Barton (1977) *Arid Land Irrigation in Developing Countries: Environmental Problems and Effects.* Oxford: Pergamon.

Author Index

Subject Index